TOPICS IN ENERGY AND RESOURCES

Studies in the Natural Sciences

A Series from the Center for Theoretical Studies
University of Miami, Coral Gables, Florida

A Continuation Order Plan is available for this series. A continuation order will bring delivery of each new volume immediately upon publication. Volumes are billed only upon actual shipment. For further information please contact the publisher.

ORBIS SCIENTIAE

TOPICS IN ENERGY AND RESOURCES

Chairman
Behram Kursunoglu

Editors
Stephan L. Mintz
Susan M. Widmayer

Scientific Secretaries
Chui-Shuen Hui
Joseph Hubbard
Joseph Malerba
George Soukup

Center for Theoretical Studies
University of Miami
Coral Gables, Florida

PLENUM PRESS • NEW YORK AND LONDON

Library of Congress Cataloging in Publication Data

Orbis Scientiae, University of Miami, 1974.
   Topics in energy and resources.

   (Studies in the natural sciences, v. 7)
   Includes bibliographical references.
   1. Power resources—Congresses. I. Mintz, Stephan, ed. II. Widmayer, Susan
M., ed. III. Miami, University of, Coral Gables, Fla. Center for Theoretical
Studies. IV. Title. V. Series.
TJ153.066   1974   333.7   74-14644
ISBN-13:978-1-4613-4537-4   e-ISBN-13:978-1-4613-4535-0
DOI: 10.1007/978-1-4613-4535-0

**Part of the Proceedings of Orbis Scientiae
held by the Center for Theoretical Studies,
University of Miami, January 7-11, 1974**

© 1974 Plenum Press, New York
Softcover reprint of the hardcover 1st edition 1974

**A Division of Plenum Publishing Corporation
227 West 17th Street, New York, N.Y. 10011**

**United Kingdom edition published by Plenum Press, London
A Division of Plenum Publishing Company, Ltd.
4a Lower John Street, London, W1R 3PD, England**

PREFACE

This volume contains the papers presented
during the Energy Session of the first Orbis
Scientiae of the Center for Theoretical Studies,
University of Miami, Coral Gables, Florida.
With this first Orbis, which met January 7-
11, 1974, the Center for Theoretical Studies
has inaugurated a new series of annual gather-
ings devoted to the natural sciences and to
problems on the "interface" of science and
society.

This volume contains, in addition, a
contribution by Behram Kursunoglu on the
"Energy Black Hole", which was not actually
presented during the Orbis.

The content of the talks presented ranged
over a wide variety of topics in the fields
of future energy needs and resources.

Special thanks are due to Mrs. Helga
Billings, Miss Sara Lesser, and Mrs. Jacquelyn
Zagursky for the typing of the manuscript
and for their efficient and cheerful attention
to the details of the conference.

THE EDITORS

CONTENTS

OPENING REMARKS

Edward Teller

Lawrence Livermore Laboratory

University of California at Livermore

Livermore, California 94550

It has been true throughout human history
that knowledge has given rise to practical results
and practical results have given rise to knowledge.
In our modern and slightly more complex terminol-
ogy we say that science contributes to technology
and technology to science.

In the new turn of human events there is a
most urgent need for more and cheaper technology
to satisfy the ever increasing demands for energy.
There is no doubt that science can make great contributions to the solution of this problem. In
turn, the solution will open up new pathways in
science.

In the present session we shall put particularly great emphasis on those solutions of the
energy crisis where scientific contributions are
expected. One of these topics, however, is ex-
cepted: solar energy will be discussed in a

separate session.

Our first discussion will be connected with
the popular source of energy which now seems to
be on its way to drying up--namely oil. The
science of oceanography suggests that the deep
portions of the ocean, in particular the continental rise, may contain tremendous additional deposits of oil. To the practical promise is closely
interwoven the problem of how to understand the
structure of ocean basins.

The much more ample deposits of solid fossil
fuels are commanding more interest due to our present difficulties. The best way from the point
of view of cost, and from the point of view of an
undesirable environment would be to derive clean
forms of fuel from coal and shale deposits by
operations contained underground. This would be
done in many ways which are being discussed. One
of the ways involves the application of nuclear
explosives for constructive purposes. This project which we call Plowshare is but another demonstration that any branch of technology can be
used for peaceful purposes, as well as for what
we call defense.

Nuclear reactors present an obviously great
promise for the rapidly expanding need of energy
in the form of electricity. Doubts have been expressed whether these nuclear reactors are safe
enough; whether they can be cheap enough, and
whether enough fuel will be available for them

in the decades and maybe centuries to which we
have to look forward. We shall hear from one of
the experts who has been involved in the process
to transform nuclear science into advanced nuclear technology.

Finally, we arrive at the frontiers of modern
technology which is controlled fusion. While the
rewards may be at a relatively distant time as
far as cheaper energy is concerned, the study of
controlled fusion has already led to an understanding of the fourth state of matter, the high
temperature plasmas. In a universe populated by
stars this state is actually the most common. In
the study of fusion we shall make many more scientific discoveries, including discoveries in astro-
physics, before the eventual technological success
is achieved and fusion becomes available to every-
body for the clean generation of electricity based
on an inexhaustible fuel.

The significance of what we shall hear and
discuss goes far beyond the solution of a world-
wide problem of great importance. It goes beyond
the question of cars lined up in front of gasoline
pumps in the United States, beyond the more worri-
some question of depression--threatening in Europe
and in Japan, even beyond the urgent threatening
propsect of starvation in countries which had prof-
ited from the green revolution and which now
experience a shortage of nitrogen based fertilizers
which are derived from gas and from oil. What we

shall discuss is of great significance to the
basic problem of linking science and technology
without which a good life for the billions on the
earth is impossible, and without which the progress of science itself will come to a stop.

Gary Higgins and Edward Teller

FUTURE ENERGY FROM FOSSIL FUELS[*]

Gary Higgins

Lawrence Livermore Laboratory

Lawrence, California   94550

It is not possible in a short paper to describe the entire fossil fuel technology in any detail. However, for completeness, many of the more commonly used recovery techniques as well as some novel but untested methods are briefly described herein.[**]

The necessity for meeting immediate and projected domestic energy demands; a rapidly decreasing ability to meet those demands from conventional domestic energy supplies; and

---

[*] This report was prepared under the auspices of the Atomic Energy Commission. For publication in January 1974 Orbis Scientiae Energy Volume.

[**] The material included in this report comes from a very large number of government and private sources. The author has made no attempt to reference these sources, so as to preserve confidentiality where that is important to future similar estimates. The opinions are his own.

recent indications that relatively cheap supplies
of foreign oil are becoming at least as costly
as domestic conventional supplies and alternatives, all focus attention on the urgent need
for developing domestic energy sources as rapidly
as possible. A considerable effort has been
devoted to R & D on intermediate and long-term
energy alternatives, but little attention has
been given to short-term alternatives.

Known domestic fossil-fuel resources include petroleum, natural gas, heavy-oil and
tar-sand deposits, oil shale, and coal. All
of the methods I shall propose are in situ
processes. These processes offer significant
environmental, ecological, economic, and aes-
thetic advantages in my judgment.

Before discussing the technology I would
like to note that supply and demand projections for 1985 are no longer correct and should
be carefully re-examined. They are based on
historical prices and industrial activity during
a period when returns on investments were most
profitable from imported oil. The world oil
market has changed during the past year, and
I expect this trend will continue. In addition,
there is every indication that the Oil Producing
and Exporting Countries (OPEC) will be reluc-
tant to increase production rates to satisfy
all demands. If average oil prices are raised
to $6.00 per barrel, domestic drilling and
finding will increase, resulting in additional

reserves and production rates. In short, I
believe that actual supplies will exceed supply projections and actual demands will be
less than demand projections. While I cannot
quantify these changes, I believe they will
not be of sufficient magnitude to relieve our
energy shortfall in the mid-1980's. Therefore, new methods and sources need investiga-
tion and development.

All of the proposed methods are designed
to produce definitive results within three to
five years after the initial year of investi-
gation. They will lead to moderately increased production by the end of the fifth
year, and should result in a significant increase by 1985, if they are started soon (1974
to 1975). Benefits will continue well into the
intermediate term (1985-2000) when other alternative sources must begin to supply an increasing
share of the domestic energy market.

What is most important is the impact that
these new technologies could have on near-term
production rates. It is estimated that their
success could result in an additional 1.5 billion barrels of oil and about 6 trillion cubic
feet of gas per year by 1985. Thus, they have
the potential of replacing about 30% of the
projected oil imports in the mid-1980's, plus
significant additions of natural gas. Estimated
wellhead costs for the oil produced is about
$5.00 per barrel, and for natural gas about

$0.50 to $1.00 per thousand cubic feet. By
1995, projected recovery rates could be about
6-1/2 billion barrels of oil and about 30 trillion cubic feet of gas per year.

The specific technologies which require
effort include: (1) Oil Recovery by Fluid Injection; (2) Oil from Heavy Oil and Tar; (3)
Oil and Gas from Stimulating Tight Formations;
(4) Oil and Gas from Coal by In Situ Conversion
of Coal; (5) Oil and Gas from Oil Shale; and
(6) Oil and Gas from Advanced Rapid Drilling
Techniques.

OIL RECOVERY FROM FLUID INJECTION

For nearly 100 years it has been common
to inject fluids to stimulate oil production
after the initial reservoir driving energy
(from a gas cap, gas in solution, or edge-
or bottom-water drive) has been depleted. Today, waterflooding is responsible for about
half of our domestic oil production. Despite
this long history, the present average efficiency of recovery of domestic petroleum deposits by all methods is only about 1/3 of
the original oil in-place, and it is estimated
that this recovery efficiency is increasing
only about 0.5 percent per year. Since each
one percentile increase in production efficiency
adds two to three billion barrels to our proved
reserves, efforts to increase this efficiency

are very important.

In the past, considerable research has been
completed on various methods of (1) waterflooding, with and without chemical additives; (2)
miscible-phase methods (in which a buffer slug
is injected that is miscible with the residual
crude oil in the reservoir, followed by a fluid
miscible with the buffer); and (3) thermal
methods. All involve fluid injection. Miscible-
phase methods include the injection of micellar
solutions, followed by polymer solutions; light
hydrocarbons; high-pressure gas in deep reservoirs; carbon dioxide; and other agents. Thermal
methods include reverse and forward in situ
combustion, steam injection (line-drive and
cyclic or "huff and puff"), and the injection
of hot gases and hot water.

Principal difficulties cited by the industry have arisen from the natural heterogeneity
of petroleum reservoirs, which causes bypassing
of entrapped residual oil and "fingering",
especially in miscible-phase recovery. It has
also been observed that those methods have
been technologically successful but economic
failures at past prices.

The research proposed in cooperation with
industry is to select tentative sites for pilot
recovery experiments; comprehensively evaluate
those sites by drilling, coring, logging, and
well testing; select the recovery method or
combination of methods best suited to the

particular petroleum reservoir conditions; conduct a pilot test and expand later to a larger
area if preliminary results are successful,
with such modifications in recovery procedures
are indicated by the pilot test.

One particularly interesting possibility
is using a carbon dioxide miscible-phase recovery
technique in a petroleum reservoir adjacent
to the site of a coal-gas synthesis plant which
will provide the needed carbon dioxide as a by-
product.

The present annual funding effort of the
United States petroleum industry in stimulative
(fluid-injection) research, from data obtained
from the American Petroleum Institute and individual petroleum companies, is estimated
at $25 to $35 million. The comparable current
annual effort of the Federal Government is
about $1.3 million, which should be increased.

Optimum applications of existing and improved methods of stimulating petroleum production by fluid injection (secondary and
tertiary recovery) can conceptually be applied
to some 60 billion barrels of oil now technologically but not economically recoverable.
Preliminary results of an increased research
effort will be available within three to five
years; the production of 400 million barrels
of oil per year should be possible by 1985,
and 800 million barrels per year by 1995.
Associated gas might be 1 TCF per year in 1985

and 2 TCF per year in 1995.

OIL FROM HEAVY OILS AND TAR SANDS

Thermal and solvent processes to recover
heavy oils have been tried with some success
but no large-scale developments have resulted.
No in situ methods have been proven for shallow
heavy oil deposits or tar-sand deposits. This
research will lead to field tests of solvent
injection-formation fracturing methods of re-
covering oil from domestic tar-sand and heavy
oil deposits. Research, now in progress, on
the use of solvent extraction for producing
heavy oils by a line-drive method in a field
test in Kansas and by a cyclic injection/produc-
tion method in California will be used to improve the technology for testing on additional
wells. Research on the characterization of
Utah tar-sands should be conducted and in situ
methods of producing the bitumen should be
developed and tested.

Ongoing research, both public and private,
at a present rate of about $2 million per year,
provides a base for an orderly expansion of
activities, leading to significant production
by 1980. Because of the relatively wide varia-
tion in resource characteristics, adaptation
of solvent and thermal methods to meet particular
conditions is needed, rather than the development of totally new extraction methods.

The objective is to develop and demonstrate efficient solvent and thermal methods of producing oil from 100 billion barrels of heavy oil and 30 billion barrels of bitumen in domestic deposits. Preliminary results will be available within three to five years; the production of 300 million barrels of oil per year should be possible by 1985, and 600 million barrels per year by 1995.

OIL AND GAS FROM STIMULATING TIGHT FORMATIONS

Hydraulic fracturing is not a new development; it has been extensively used for the stimulation of oil and gas wells for the past 20 years. However, it has not been attempted on the scale now contemplated for the low permeability deposits of the western states.

High explosives were used for well stimulation in the early years of the oil industry but were replaced by hydraulic fracturing because of the inherent danger of chemical explosives and because hydraulic fracturing gave improved results. It now appears possible to combine hydraulic fracturing with the injection of new, less dangerous liquid explosives into the formation to take advantage of both techniques. While some explosives for downhole emplacement have been developed, they have not been adequately tested under actual field conditions.

Field tests of massive hydraulic fracturing (MHF), high explosive fracturing, and a combination of the two are needed to develop and optimize the fracturing techniques.

It is expected that these methods will not work successfully in every geologic environment. The procedure will have to be modified to meet local conditions. This requires tests at a number of sites with different geologic conditions.

Nuclear stimulation has been shown to be technically feasible by three field experiments conducted in the past few years, including one using multiple explosions. Results of these tests indicate increases in potential 20-year production of a factor of 3-15 over conventional techniques. It is expected that massive hydraulic fracturing, high explosive fracturing, combination hydraulic/explosive fracturing, and nuclear explosive fracturing will prove to be complementary techniques, each with certain advantages and limitations depending on the resource locations. In each case a developing technology can be accelerated to a commercial level by 1979. The techniques are expected to lead to a combined gas-production potential of more than 3 trillion cubic feet per year by 1985 and 10 trillion cubic feet per year by 1995. The projected increase in oil production is less easily defined, but could be 100 million barrels per year by 1985 and 200 million barrels per year by 1995.

OIL AND GAS FROM COAL

The fundamental R & D for all three elements
of this effort (high-Btu, low-Btu gas and methyl
fuel) has been underway for decades and should
be completed in 1975. There have been contributions from government and private sectors
in both the U.S. and almost all European countries. Pilot plant tests should be finished
by 1978 or 1979. First commercial scale operations are expected by 1980 to 1982 and, because
the unit prices, capital investment, and profit
margins are very favorable, these technologies
are expected to be rapidly commercialized.

Coal is our most abundant fossil fuel resource and $20,000 \times 10^{15}$ Btu of fuel value
should ultimately be recoverable by these technologies (290 years energy supply at the 1970
rate). Therefore, there should not be a resource base limitation in the near future.

Possibly the most serious technical barrier
to implementation is varying success in different coal deposits in different locations.
Thus, there will always be a risk in each new
location.

Low-Btu gasification is now in the pilot
plant stage so both it and methyl fuel production are virtually certain of technical success.
High-Btu gasification is technologically more

difficult and is more site dependent.   The major
problems are finding the right sites, proof of
the cost estimates, and engineering demonstration of process and quality control.

     Several different developing fracturing
techniques and chemical processes involving
partial in situ combustion should be accelerated.   Each of these should be tested
and the more favorable brought to commercial
production with a projected annual rate of 0.3
trillion cubic feet of gas and 20 million barrels of oil equivalent by 1985, and 4 trillion
cubic feet of gas and 0.5 billion barrels of
oil equivalent by 1995.

OIL AND GAS FROM OIL SHALE

     The objective of this research effort is
to develop appropriate fracturing techniques
(for example, nuclear explosives, chemical
explosives, partial mining, hydraulic fracturing,
or combinations thereof) and processing methods
(for example, in situ combustion, circulation
of hot gases, or hydrogasification) for the
in situ production of oil and gas from oil
shale.

     Interest in in situ processing oil shale
in the last 20 years or so has been largely
in the U.S.A., but research has been sporadic
until the last few years, during which the
government has funded a program.   The program,

which is composed of laboratory and small field
experiments, has investigated hydraulic fracturing
and evaluated nuclear explosives, chemical
explosives, and electrical energy as agents
for fracturing the oil shale to achieve the
permeability required for the passage of gases
and liquids in the recovery process. In situ
combustion and forward drive have been used
to try to retort the shale bed. All of these
methods are in the early stages of development.

Within the past year an industrial company
has undertaken a small scale in situ experiment in which some of the shale is removed
by mining and the remaining shale in the bed
broken by chemical explosives followed by forward burning retorting. They report optimistic
future development of this method. All research
should lead to a production of 200 million
barrels of oil and 0.1 trillion cubic feet of
gas per year by 1985 and 2 billion barrels of
oil and 1 trillion cubic feet of gas per year
by 1995.

OIL AND GAS FROM ADVANCED DRILLING TECHNOLOGY

The largest cost and the most time consuming operation in discovering and developing
an oil or gas well, particularly a deep well,
is the drilling cost. Cheaper and faster drilling methods will increase the exploration

rate, and thus the number of finds, and the rate
of development of the finds. New oil is found
in only 10-12% of exploration wells.

        One technique currently under investigation
by industry, is hydraulic jet drilling. In
this method, fluid, with or without abrasives,
is expelled from high pressure nozzles at the
drill bit face to erode the rock. This method
has been tested at relatively shallow depths,
but further R & D is needed, particularly to
develop equipment capable of handling the high
pressures and long life required.

        A second method is the use of the cavitation
effect of pulsed electrical detonations. This
technique, called spark drilling, has been
field tested with some success in the USSR,
but only at bench scale in the U.S. The
limiting factor in the USSR tests appears to
have been the need for a downhole pulse generator,
which we believe we can develop.

        The objective is to develop methods to drill
cheaply and rapidly under conditions at which
present drilling methods approach their economic
limits. While it is extremely difficult to
estimate the impact of successful development
of these techniques, it can be conservatively
estimated from National Petroleum Council data
that a 3-1/2% increase in drilling activity,
primarily in deep reservoirs, could yield up to
0.5×10^9 bbl/year of new oil and 2.5 TCF/yr
of new gas by 1985. Since most of this oil and

ADVANCED METHODS OF OIL AND GAS RECOVERY FROM FOSSIL FUELS (IN SITU)

Estimated Annual Additional Oil and Gas Production Rates

	1985			1995		
	OIL		GAS	OIL		GAS
	Billion Barrel	10^{15} Btu	10^{15} Btu or TCF	Billion Barrel	10^{15} Btu	10^{15} Btu or TCF
1. Fluid Injection	0.4	2.3	1	0.8	4.7	2
2. Tar & Heavy Oils	0.3	1.8	–	0.6	3.5	–
3. Explosive Stim.	0.1	0.6	3	0.2	1.2	10
4. Coal	0.03	0.2	0.3	0.5	3	4
5. Shale	0.2	1.2	0.1	2	11.7	1
6. Advanced Drilling	0.5	2.9	2.5	2.5	14.6	12.5
TOTALS	1.5	9.0	6.9	6.6	38.7	29.3

Total Oil & Gas (10^{15} Btu/y)   15.9   68.2

Estimated Wellhead Prices

	Oil ($/bbl)	Gas (¢/Mcf)
1. Fluid Injection	5	80-100
2. Tar & Heavy Oils	6	-
3. Stimulation	4	40-100
4. Coal	3-6	40-60
5. Shale	4-5	50-100
6. Drilling	4-5	40-100

Assumptions:

1. Federal leasing is pursued vigorously - both off-shore and for oil shale (affects lines 5 and 6).

2. Nuclear fracturing of tight gas sands and oil shale is permitted. (Affects lines 3 and 5.)

gas has not yet been discovered, the probability
of attaining this large potential payoff is not
as high as the probability of success in the other
research areas where the target resource is
known. While the potential oil and gas yield
of advanced drilling may be initially large,
it must be realized that the long-term effect
would be to speed eventual depletion of oil
and gas reserves. Improved drilling must therefore be viewed as a stop-gap measure which can
buy time for implementation of alternate energy
sources.

        The table following summarizes the conse-
quences of all of the research and development
described above, and includes estimates of the
prices of some of the products (in 1972 dollars).

SOLAR ENERGY, ITS CONVERSION AND UTILIZATION

Erich A. Farber, Director

Solar Energy and Energy Conversion Lab.

University of Florida

Gainesville, Florida

ABSTRACT

This paper discusses the part solar energy, our only energy income, can play in meeting our ever increasing energy demands. The basic characteristics of solar energy and the solar properties of materials are mentioned. The conversion of solar energy to the forms of energy which are needed in our daily life are explained.

The basis of the discussions is the University of Florida Solar Energy and Energy Conversion Laboratory with its Solar House and its Solar-Electric Car.

INTRODUCTION

It is almost impossible to pick up a newspaper or magazine today without being reminded of

our present energy situation which by many is referred to as a "crisis".

Our present civilization has been built upon
the energy provided by fossil fuels (coal, oil and
gas), resources which have been provided for us by
nature through rather inefficient conversion of
solar energy. This is nature's way of storing
energy through plant life and just the right conditions of oxygen starvation, high pressure and
temperature coupled with millions of years of time.

This energy source can be compared with the
savings account of a family. Knowing how much is
consumed each day, simple arithmetic will reveal
how long the family can survive drawing upon their
savings, or how long civilization can live as it
has become accustomed to before running out of
these energy deposits.

Everyone agrees with the above, only differ-
ences of opinion arise from the question as to "when'
this will occur. Not much is gained by arguing
about the exact time since we do not know how much
of these resources exist - we can only estimate -
and no one can predict how sensibly and effectively we are going to utilize them.

One further observation should be made, namely that these fossil fuels are also the source of
our chemical industry, our medicines, preserva-
tives, etc., and if they are burned for energy
they will be unavailable for those other uses
equally essential for life.

Not much energy has been used per person for

thousands of years until about 100 years ago.
With the increase of energy consumption each person has effectively multiplied the number of "ser-
vants" working for him and so quantities of goods
and the standard of living increased rapidly. The
richer a country is in energy and the more it uses
the faster is its climb to the heights aspired by
mankind as expressed by its standard of living.

The well-being of civilizations will parallel
the availability and consumption of energy. With
fossil fuels being used at a tremendous rate and
becoming scarcer, we will have to curtail our
activities and possibly even accept a decline, or
find other sources and methods to provide for our
energy needs and requirements. We will have to
develop nuclear energy and especially to learn to
live off solar energy, our only energy income, to
conserve our energy savings.

Because of the above, about two decades ago,
the University of Florida Solar Energy and Energy
Conversion Laboratory was established with the objective to study the feasibility of providing the
forms of energy needed for our daily life by con-
verting solar energy. The conversion is to be done
in the fewest possible steps along the most direct
route. The activities of the laboratory cover:
the monitoring of solar radiation, the determination of solar properties of materials, solar water
heating, solar swimming pool heating, house heating,
air-conditioning and refrigeration, solar cooking
and baking, solar distillation, high temperature

applications - solar furnaces, solar power generation both mechanical and electrical, solar sewage
treatment, liquid waste recycling, solar-electric
transportation, etc.

The laboratory has a solar house where many
of the devices developed in the laboratory are
being used and their performance is observed. It
also has a solar-electric car which is driven by
a staff member daily to obtain "in use" performance
information. In the following pages, some of the
larger projects are discussed.

SOLAR ENERGY

As was mentioned above fossil fuels which are
really stored solar energy will have to be supple-
mented by other energy resources, especially by
the direct use of solar energy, the output from an
old, large and safe nuclear plant.

Solar energy is the only energy income we
have, it is well distributed, free and does not
add anything degrading to the environment through
its conversion and utilization. The end product
is heat whether conversion steps are interposed or
not. It arrives at about one horsepower per
square yard, or several times the amount of energy
falls on the roof of a typical house as could
possibly be used inside.

Since solar energy is intermittent, only
available during the day, for night time operation,
if required, storage or another source must be

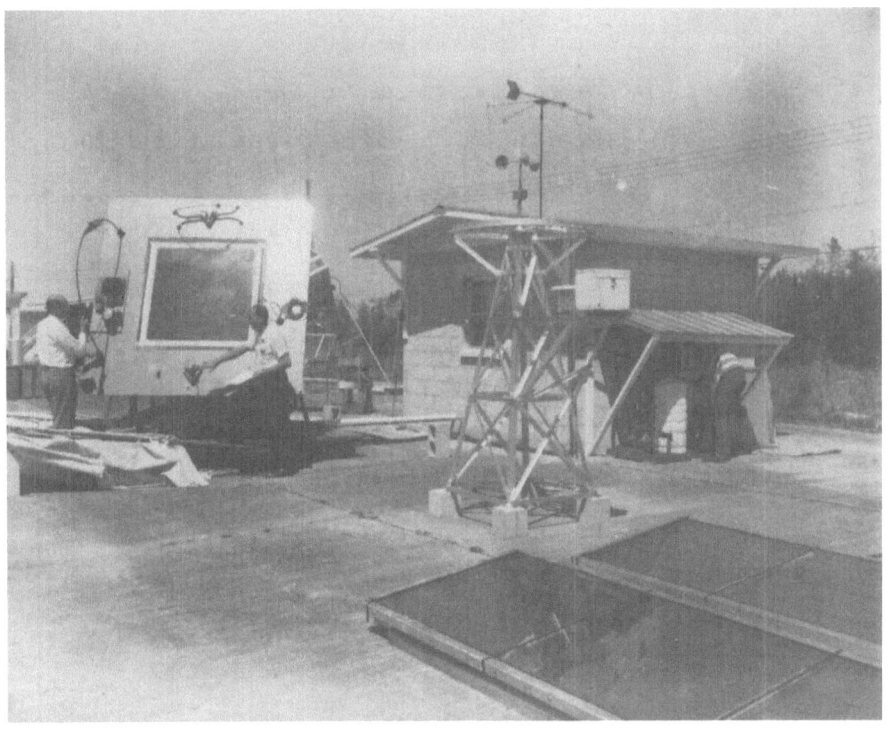

Figure 1

provided. Technically it is possible today to
convert solar energy to all the forms of energy
needed. Some of these conversion methods are more
competitive economically than others.

Fossil fuels and nuclear fuels can best and
most efficiently be converted in large central
stations. Even though large conversion units have
been proposed for solar energy, it is questionable
whether its basic characteristics match these
schemes. It might be better to convert solar
energy in small units at the location of use rather
than collect it somewhere else, convert it, and
then transport it to locations where it is found
in the first place.

About 100 stations in the United States moni-
tor the amount of incoming solar radiation continuously providing the necessary information for
utilization at a specific location.

SOLAR PROPERTIES OF MATERIALS

The University of Florida Solar Energy and
Energy Conversion Laboratory has extensive facilities to study the solar properties of materials.
The most flexible instrument is the solar calorimeter which provides almost all the information on
solar characteristics of fenestrations published
in the GUIDE of the American Society of Heating
and Air-Conditioning Engineers, which is used extensively by all architects and engineers, Figure 1. It is essentially a well instrumented

black cavity which allows the mounting of materials
to be studied as shown. Extensive instrumentation
allows the measurement of incoming energy of high
and low wave length radiation, convection and con-
duction, and delicate heat balances provided by
hundreds of thermocouples. The instrument can be
oriented in any desired position with respect to
the sun, simulating walls and inclined or flat
roofs, or it can be made to follow the sun.

A small weather station and an instrument
building go with the calorimeter. It is possible
to simulate winter conditions by use of a refrigeration system, summer conditions by means of electric heaters, or ambient conditions if desired.
Utilizing the real sunshine rather than artificial
sources has been found of considerable value since
simulations did not seem to be very reliable.
Blowers on the instrument can simulate wind conditions for the tests. Photospectrometers, hot
and cold boxes, etc., are also available.

Materials such as glasses (plain, tinted,
coated, laminated, multilayered), plastics (transparent and translucent), glass brick, venetian
blinds (some of them water cooled), drapery materials, sun screens of all kinds, etc. have been
investigated. (2,3,4,5)

In addition to the determination of solar
properties, the laboratory has exposure test facilities to evaluate the weathering properties of
materials. This is often done before studying the
materials in detail. Two 5 foot diameter solar

furnaces are used for high temperature work.

Once the true properties of materials are
known, the best can be selected for the conversion
of solar energy into the required forms.

SOLAR REFRIGERATION AND AIR-CONDITIONING

One of the real needs is the ability to
preserve food. This can be done by solar refrigeration which is ideally matched to the energy
supply, cooling is needed most when the sun shines
hottest. Solar energy can be used to drive an
engine and then a compressor to provide compression
refrigeration, or the heat from solar energy can
be used in an absorption refrigeration system.
These and a few other systems have been developed
in the laboratory.

Steam jet refrigeration was tried at one time,
and oil heated by the sun was used to replace the
gas flame of a gas refrigerator. These methods
worked but were not considered the best since they
required concentration and thus could not utilize
the diffuse portion of the solar energy and would
not work on cloudy days.

For this reason the emphasis was put on flat
plate solar collectors providing the energy to
operate absorption refrigeration and air-conditioning systems. Flat plate solar absorbers heat
water which is the energy source for absorption
refrigeration or air-conditioning systems. In an
ammonia-water system the heat drives the ammonia

Figure   2

Figure   3

from solution. The ammonia is then condensed and
the liquid expanded. This makes it very cold and
able to absorb heat, thus providing the cooling.
The warmed vapor is reabsorbed in the water and
circulated back to start its cycle over again.
The process can be carried out intermittently or
continuously.

Figure 2 shows a 5 ton air-conditioning system, Figure 3 a small solar ice machine which can
produce as much as 80 lbs. of ice on a good day.
(2,7,8)

Storage can be provided in the form of hot
water, ammonia, or ice. The latter has the advantage that it can be moved to different locations and therefore service other than just the
immediate area. The ice machine has a 4 ft. x
4 ft. solar collector which serves at the same
time as the ammonia generator, not requiring solar
water heaters. Its conversion is slightly better
since no heat is lost in heat exchangers.

A water driven air-conditioning system is the
easiest to combine with a solar heating system
permitting double use of many parts.

SOLAR POWER GENERATION

A rather extensive program in the laboratory
deals with power generation. Many engines of
different designs and operating on different
principles have been designed and evaluated. Some
of them do not have moving parts. However, at

Figure 4

Figure 5

this time it seems that the vapor and hot air
engines have the most promise. A number of frac-
tional horsepower vapor engines and hot air engines have been built and used to pump water, drive
machinery, or drive electric generators to charge
batteries for night use or for transportation in
the solar-electric car.

The closed cycle hot air engine shown in
Figure 4 can develop about 1/3 horsepower, the
limitation not being the engine but the concentrating mirror which is about 5 ft. in diameter. A
larger mirror would allow the engine to put out
more power. These engines only need a source of
heat and therefore can be operated with wood, coal,
gas or oil, if night time operation is required.
In the closed cycle hot air engine the enclosed
air is alternately heated and cooled when brought
in contact with the hot and then the cool walls.
When the air is hot the pressure is high and the
power piston is pushed down; when the air is cool
the flywheel returns the power piston against low
pressure. A plunger moves the air back and forth
between the hot and cool walls. (2,9,10) In the
closed cycle engine the speed of the engine is
controlled by how fast the air can be heated and
cooled. To separate the speed of the engine from
the heat transfer characteristics open cycle
engines were designed. Figure 5 pictures one of
those engines.

In the open cycle hot air engine the air is
taken in and compressed. It is then moved through

Figure 6

a heater where it reaches high temperatures. From
the heater it flows through the engine where it is
expanded, doing work, and then exhausted to the
atmosphere. The engine and the compressor are
coupled together. By this method the engine
speed and the heat transfer characteristics are
independent.

The above two engines require concentration
of solar energy and thus need rather good days for
operation. Vapor engines have been designed and
built which use flat plate absorbers to generate
vapor at relatively low temperatures and use it in
vapor cycles.

SEWAGE TREATMENT

Among other applications of solar energy is
the treatment of sewage. It was found that solar
energy can be used to keep the sewage digester
temperature up to provide more efficient bacterial
activity. In this manner the sewage handling
capacity of digesters can be considerably increased.
Figure 6 pictures solarly heated sewage digesters
which were used in the study.

THE UNIVERSITY OF FLORIDA SOLAR HOUSE

Approximately seven years ago the conversion
of the University of Florida Test House into a solar
house was started. This was done step by step as
time and funds permitted.

The reasons for utilizing this house were many
and among the most important ones were that information existed over the last 13 years as to how
different conventional systems performed in supplying the hot water, heating, air-conditioning, the
energy for cooking and other activities while monitoring the air quality in the house. All the data
in this thoroughly instrumented house was taken
while a married student couple lived in this house
with all modern conveniences provided.

In the early stages of the project, walls,
windows, and insulation in the house were changed
to evaluate their performance and later the systems serving the house were evaluated i.e.; oil,
gas, and electricity were used at different times,
to provide the energy to water heaters, air-con-
ditioners, heat pumps, cooking systems, etc. Over-
head or attic air distribution systems with different diffuser outlets were compared with under
floor distribution systems.

These years of data and experience, under
actual lived-in conditions, give a wealth of information and a firm basis for absolute compari-
sons of different systems serving the same house
under the same conditions.

The house is a conventional, typical block
construction dwelling similar to many found in
Florida and elsewhere. It has three bedrooms,
two baths, kitchen, living and dining rooms,
utility and laundry rooms, and a carport with
closed in storage space. The laundry room,

Figure   7

besides holding a washing machine and dryer, is
used for all the instrumentation monitoring the
many activities and systems of the house.

Figure 7 shows a SW view of the house which
is oriented E to W. The road approaching the
house comes from the E so the solar equipment is
not seen until one walks around the house. All
the solar energy equipment which is installed in
the solar house was developed and evaluated in the
University of Florida Solar Energy Conversion
Laboratory.

The first unit which was added to the house
was the solar water heater. The collector was put
on the roof with the hot water tank behind it.
The first visitors to the solar house were dis-
appointed that they did not see the solar equipment, and it was too hazardous to take them up on
the roof. For educational reasons it was decided
to place the rest of the solar energy conversion
equipment next to the house in the open, rather
than to incorporate it in it, so that visitors can
walk around the equipment, touch and photograph
it and in general get a good idea what such equipment is like.

Further, two swimming pools were added, one
heated by solar energy and the other as standard
for comparison; a house heating system with a large
storage tank, above ground rather than buried,
with plans to use the solar collecting and storage
parts also for the air-conditioning; a small
liquid waste recycling system; a small solar

Figure   8

Figure 9

energy to electricity conversion unit; and a solar-
electric car which is part of the overall system.
In the near future, solar air-conditioning, refrigeration and cooking will be added.

THE SOLAR WATER HEATER

The solar heater is shown in Figure 8. It
consists of a 48 ft^2 solar collector and a 80 gal.
well insulated hot water storage tank. (3,13,14,
15,16,17)
The solar absorber is a galvanized sheet metal
box having 1 inch of styrofoam insulation inside
in the back. In front of the insulation is a
copper sheet with two parallel circuits of sin-
usiodally arranged tubes soldered onto it.
sheet and tubes are painted with a good absorbing
paint. The box is covered by glass having good
solar energy transmitting properties.
The hot water delivered by this unit flows
by free convection to the hot water storage tank
which is well insulated to reduce heat losses.

THE SOLARLY HEATED SWIMMING POOL

To truly evaluate the effectiveness of
heating the swimming pool by solar energy, two
identical pools were installed, 15 feet in diameter and 4 feet deep. One was heated by various
methods utilizing solar energy and the other was
used as standard for comparison, Figure 9.

Both pools were well instrumented with many
thermocouples in each. One pool, the unheated one,
was left to itself. The other was heated by solar
energy in a number of ways. The simplest method
was to float a transparent sheet of plastic on its
surface. Two sheets of the air-mattress design
do a better job, or bubble sheet can be a reasonably good collector and a good inhibitor to heat
losses. The solar absorbers of the house heating
system described below were also used at times to
heat one of the pools. The latter system is economical only as a combination between house and
pool heating.

The simple plastic sheet could keep the
average pool temperature 20 F above the average
air temperature and 10 degrees F above the unheated pool temperature. Utilizing the house
heating absorbers, the pool temperature could be
kept about 40 degrees F above the average ambient
air temperature.

THE SOLAR HOUSE HEATING SYSTEM

The solar house heating system is basically
a hot water system which was selected over the air
heating system, since the former is easier to use
as the front end of a solar air-conditioning system. Ten solar absorbers, similar to the one
used for the solar hot water system comprising
350 ft^2 of absorbing surface provide hot water
which is stored in a 3,000 gallon tank with 4

Figure   10

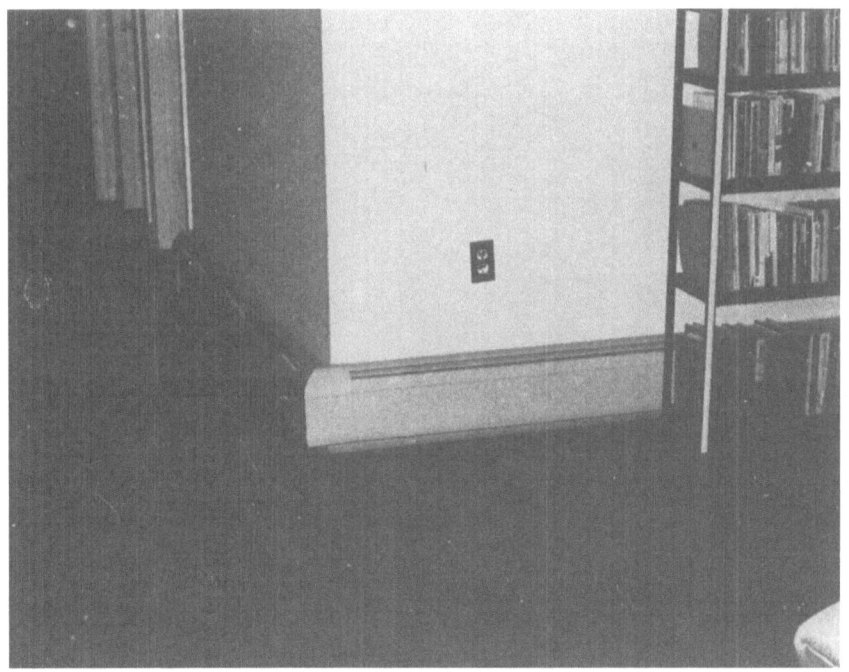

Figure 11

Figure 12

inches of insulation around it, Figure 10.

Water from the storage tank is circulated by a small pump through the baseboard heating system in the house as required to keep the temperature of the house at the desired value. 140 feet of baseboard heaters can deliver 60,000 Btu per hour with supply water of 130 F which is the design load for the house to meet the maximum heat requirement under extreme conditions in Gainesville, Florida. With the water hotter, more heat can be delivered and cycling controls the actual amount of heat delivered.

The baseboard heaters are shown in Figure 11. The flow rate through the solar absorbers can be controlled, so as to deliver water at the desired temperature, storing it in the upper part of the storage tank. The delivery to the house is thermostatically controlled.

The storage tank is larger than actually needed but was used to allow long time storage to carry the house through bad weather conditions. 25,000 Btu can be delivered to the house for only 1 F water temperature drop in the storage tank.

LIQUID WASTE SOLAR RECYCLING PLANT

Since fresh water is becoming more and more difficult to obtain and is also getting more expensive, a small liquid waste solar recycling plant has been added to the house. This solar distillation unit, Figure 12, has a liquid holding

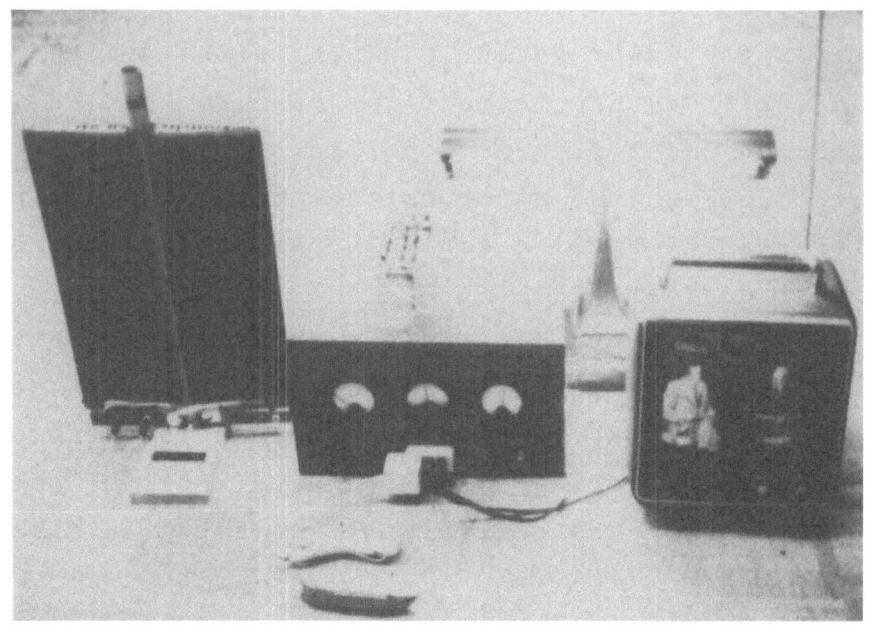

Figure   13

tray area of 22 ft^2 and can produce up to about 3
gallons of fresh water on a good day. This unit
is also designed to collect rain water which, in
Gainesville, just about doubles the output. (18)

This plant is not able to handle all the
liquid waste of the house, but one could be built
any desired size depending upon the recycling requirements.

SOLAR-ELECTRIC CONVERSION UNIT

Most of the energy today in a house is used
for water heating, house heating, and air-conditioning. The real need for electricity is only
a small fraction of the total energy requirement.
It is really only needed for radio, T.V., lights,
and some small appliances. Figure 13 shows the
small unit used to convert solar energy, by means
of solar cells, into DC electricity and store it
in NiCd batteries. The energy from the batteries
is then converted as needed by a DC to AC solid
state converter to operate lights, radios, T.V.,
and small appliances. The cost of this unit is
certainly not competitive at this time, but it
demonstrates the feasibility of providing electricity.

SOLAR AIR-CONDITIONING, COOKING AND REFRIGERATION

When the solar house heating system was designed, it was done so that the solar absorbers

and the storage system can be used to drive a
specially designed air-conditioning system during
the cooling season. The hot water is used in the
winter for heating the house and in the summer for
air-conditioning. Systems described earlier have
been designed in the University of Florida Solar
Energy and Energy Conversion Laboratory which can
be used in this manner. (7,8,19,20) The systems
operate with water as low as 120 F.

     The air distribution system to be used with
the solar air-conditioning system is already in
the house, so only the absorption system has to
be added. It is being designed so as to fit the
needs of the house. It is planned to add air-
conditioning as the next step.

     After the air-conditioning is incorporated
into the solar house, a concentrator from the
Solar Energy Laboratory will be moved to the house
to provide oil at high temperature which will be
stored in a tank. This oil will then, as needed,
be circulated around burners of a stove and in an
oven so as to allow cooking very similarly as with
an electric stove. The electric elements are replaced by coils of copper tubing.

     Such an experimental system was operated a
number of years ago in our Solar Energy and Energy
Conversion Laboratory. At that time, the hot oil
was also used to operate a refrigerator in which
the gas flame was replaced by a hot oil bath. A
better and more effective solar refrigerator has,
however, been developed since.

Figure 14

Figure 15

THE SOLAR-ELECTRIC CAR

The solar electric car of the laboratory, Figures 14,15,16 which can go about 60 miles/hour and has a range of about 100 miles is part of a system providing the energy requirements for a family, both in the house and for necessary transportation. All this comes from solar energy directly and is pollution free.

For transportation solar energy is converted at the present time, into mechanical work by a hot air engine which in turn drives an automobile generator which can charge the batteries of the electric car. This type of conversion from solar energy to electricity is much less expensive than the use of solid state conversion (12).

A network of "filling stations" each having such conversion systems and using them to charge up banks of batteries could, instead of filling the gas tank of a car with gasoline, exchange run down batteries for charged ones to provide the needs of the traveler.

OUTLOOK FOR THE FUTURE

The fact that fossil fuels, nuclear fuels, etc. are fixed in quantity may in itself not be the limiting factor in their utilization. Pollution, inherent in the conversion of these fuels may instead force the curtailment of their use. In the Los Angeles basin, limitations have been

put upon the building of new power plants. The
only alternative in this case may be to limit the
consumption per person. "Rolling Black-outs" have
been suggested as one solution. The basin can be
divided into regions and the power to each region
can be cut for one hour out of a 24 hour period,
moving consecutively from region to region.

Gas rationing has been suggested for the Los
Angeles basin during the last part of 1972 to reduce the pollution and smog problem. This limiting
of the energy supply is not due to shortage or un-
availability of fuel but due to the consequences
of using it.

Solar energy is the only large source of
energy which can be pollution-free. Its use does
not add or subtract anything to the environment,
it only interposes additional steps into the conversion from solar energy to finally heat, which
is the end product even if it is not used. This
can be likened to a river which on its way from
the mountains to the ocean drives a water wheel
providing useful energy. The water flows from the
mountains to the ocean whether energy is extracted
or not.

In summarizing it may be stated that, in this
writer's opinion, we will always need fossil fuels
for certain applications but since they will become
scarcer and more expensive other fuel sources will
have to be drawn upon if we wish to continue to
enjoy the standard of living which we are enjoying
now and if the rest of the world is to raise its

standard of living.

So we will move from a "Fossil Fuel Society" which we are at present to an interim "Nuclear Society" when a considerable portion of our energy requirements will be met by this source. But since the above sources of energy have to be classified as savings they cannot last forever and society and civilization will have to move toward a source which is well distributed, pollution-free, and inexhaustible, thus classified as income. So ultimately we will, by necessity, become a "Solar Society".

When all this will come about will depend upon our actions, wisely or unwisely, dictated in many cases by economic and political decisions.

58   ERICH A. FARBER

REFERENCES

1. C.W. Pennington, University of Florida-ASHRAE Solar Calorimeter, ASHRAE Journal, Vol. 8 No. 3, March, 1966.
2. E.A. Farber, Solar Energy: Conversion and Utilization, Building Systems Design, June, 1972.
3. E.A. Farber, Selective Surfaces and Solar Absorbers, Journal for Applied Solar Energy, April, 1959.
4. E.A. Farber, Theoretical Effective Reflectivities, Absorptivities, and Transmissivities of Draperies as a Function of Geometric Configuration", Solar Energy, Vol. VII, No. 4, Oct.-Dec. 1963.
5. E.A. Farber, Experimental Analysis of Solar Heat Gain Through Insulating Glass with Indoor Shading", ASHRAE Journal, Feb. 1964.
6. E.A. Farber, "Crystals of High Temperature Materials Produced in the Solar Furnace", Solar Energy, Vol. VIII, No. 1 Jan.-Mar., 1964.
7. E.A. Farber, et. al, Operation and Performance of the University of Florida Solar Air-Conditioning System, Solar Energy, Vol. X, No.2, April-June, 1966.
8. E.A. Farber, A Compact Solar Refrigeration System, ASME Paper #70WA/SOL4, Dec. 1970.
9. E.A. Farber, et. al, Closed Cycle Hot Air Engines, Solar Energy, Vol. IX. No. 4, Oct.-Dec. 1965.

10. E.A. Farber, Hot Air Engines, Mark's Mechani-
     cal Engineers' Handbook, 7th Ed., McGraw-Hill
     Book Co., New York, 1966.

11. E.A. Farber, "The Application of Solar Energy
     to Sewage Digestion and Liquid Waste Re-
     cycling," Proceedings of the Third National
     Convention of the Institute of Plumbing,
     Australia, March, 1973.

12. H.R.A. Schaeper, et. al., The University of
     Florida Solar-Electric Automobile, Mechanical
     Engineering, Nov., 1972.

13. E.A. Farber, Practical Applications of Solar
     Energy, Consulting Engineer, Sept. 1956.

14. E.A. Farber, Solar Water Heating; Present
     Practices and Installations, ASME Paper
     #57-SA-45, June 1957, and also National
     Engineer, Aug. 1957.

15. E.A. Farber, et al., Solar Energy to Supply
     Service Hot Water, Air-Conditioning, Heating
     and Ventilating, Oct. 1957.

16. E.A. Farber, Solar Water Heating and Space
     Heating in Florida, The Journal of Solar
     Energy Science and Engineering, Vol. III, No.
     3, Oct. 1959.

17. E.A. Farber, The Use of Solar Energy for
     Heating Water, U.N. Conference Proceedings
     on New Sources of Energy, Aug. 1961.

18. C.R. Garrett, et al., Performance of a
     Solar Still, ASME paper, Dec. 1961.

19. E.A. Farber, Solar Water Heating, Space
     Heating and Cooling, Journal of Applied

Solar Energy, Aug. 1960.

20. M. Eisenstadt, Tests Prove Feasibility of
     Solar Air-Conditioning, Heating, Piping and
     Air-Conditioning, June 1960.

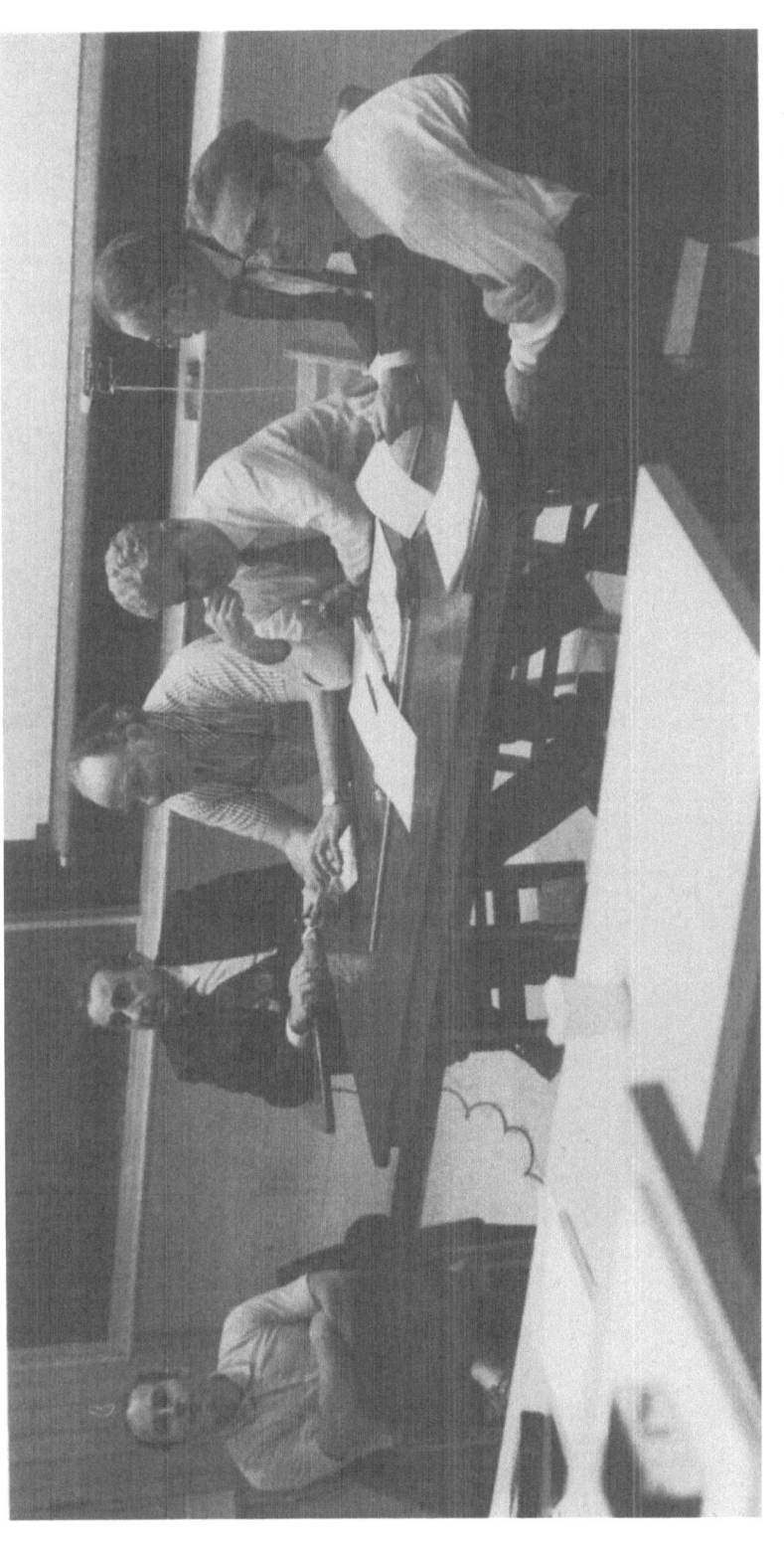

ENERGY SPEAKERS: From left: Anthony Colleraine; Edward Teller; Melvin A. Gottlieb; J. Lamar Worzel; Gary Higgins; and Henry Hurwitz, Jr. (Erich A. Farber and Behram Kursunoglu missing from picture.)

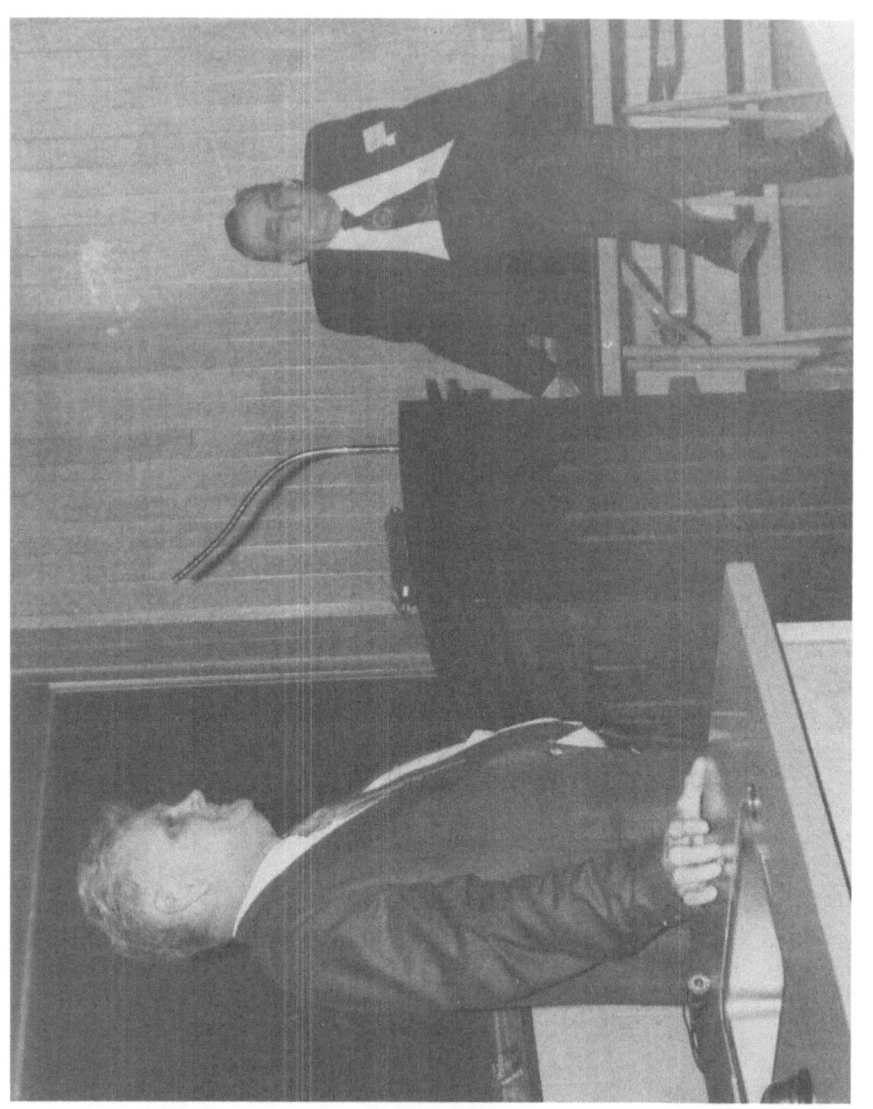

J. Lamar Worzel and Edward Teller

MARGINAL DEEP WATER AREAS OF THE USA, POTENTIAL TO ALLEVIATE THE ENERGY CRISIS

J. Lamar Worzel

Marine Biomedical Institute

University of Texas Medical Branch

Metamorphic processes in a non-oxydizing environment converts organic matter in sedimentary rocks into hydrocarbons. Whether this process is a biological, chemical or physical process or combinations of these is still the subject of continuing basic research.

Nearly all sediments are laid down in water covered areas, consequently, in these cases, all pore spaces are initially filled with water. Further sedimentation causes burial and consolidation of the sediments with expulsion of pore water. Gas and liquid hydrocarbons generated within the sedimentary layers seek higher elevations, when permeability permits, with gas on top, petroleum liquids underneath and water at the bottom (Figure 1). Sometimes gas is overpressured due to moving fluids within layers, additional consolidation of the layer or generation of additional gas within the layer when it is trapped

Figure 1.   Typical anticlinal structure showing relative locations of gas,
oil and water accumulations beneath an impermeable capping layer.
(After API)

beneath an impermeable layer. In such cases, drilling into hydrocarbon reservoirs can result in "blowouts" unless precautions are taken. The typical movie scene of an oil geyser is in reality a blowout.

Because of the mobility of fluids, petroleum is often found in beds other than the source beds in which it was formed. On a few occasions, it has even been found in highly fractured igneous rocks.

Geophysical techniques, principally magnetic, gravity and seismic methods, are used to locate and detail 1) sedimentary volumes, 2) anticlinal traps, 3) reefs, 4) faults and 5) stratigraphic traps. It is true that these features may exist without entrapped petroleum, but when source beds are present and there is an impermeable roof for the structure there is a reasonably high probability that petroleum products will be present in some of these features.

Salt Domes are formed by solid flowage of salt. Normally it rises, usually in a conical shape, into overlying sedimentary beds because it has a lower density than that of the surrounding sediments at depths exceeding a few thousand feet. At least in some cases these penetrate intrusively into overlying beds as burial proceeds. All of the above types of traps can be illustrated by structures around salt domes, as in Figure 2. The elevated ocean floor of the sea can be the location of reefal material which on later burial becomes a

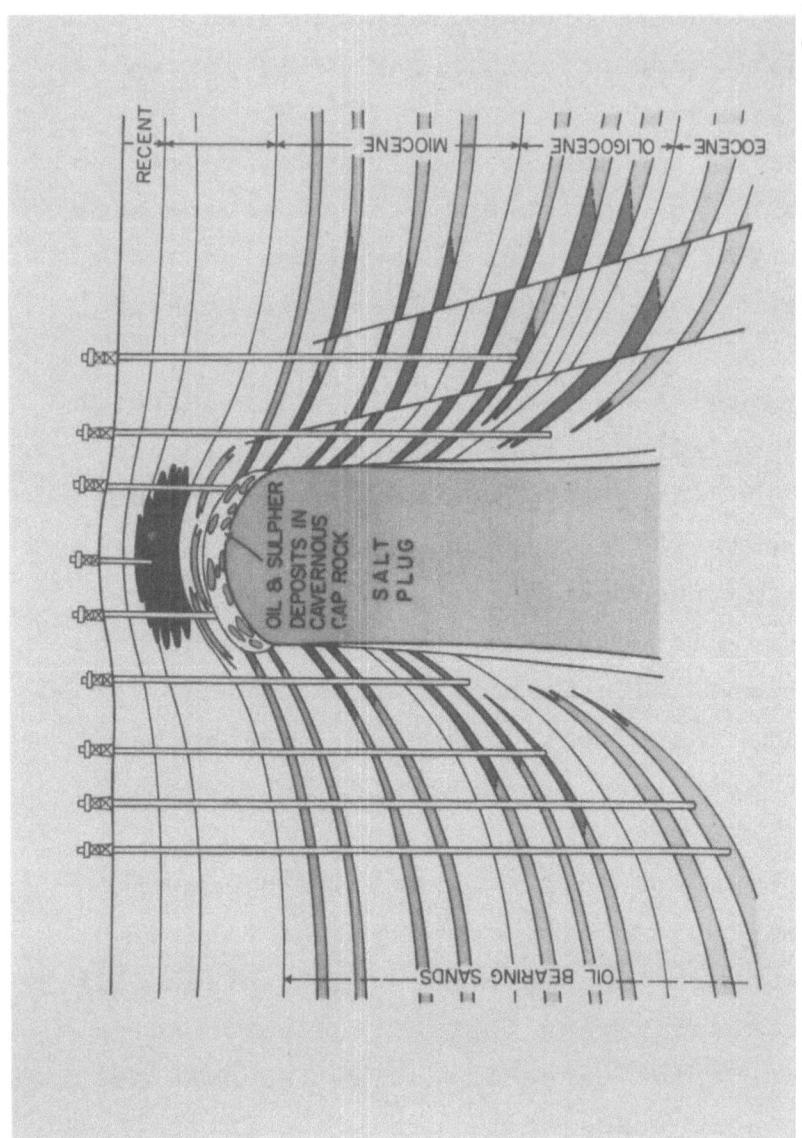

Figure 2. Illustrative diagram of petroleum traps in the vicinity of salt
domes. Note the 3:1 vertical exaggeration which makes the slop-
ing sides of the dome look almost vertical. The darkened areas
represent "oil traps" which sometimes occur in such locations.
The very dark area above the cap rock represents a reef with oil
entrapped.

special form of stratigraphic trap. The arching
of layers over a dome cause anticlines which can
trap petroleum fluids. The faults, over the top
of a dome and along the margins of a dome can seal
off permeable beds producing traps. The high density cap rock often formed just above the salt,
believed to be formed by salt solution and/or
chemical reactions replacing salt with vuggy dolo-
mite, anhydrite sulfur and gypsum, can be considered to be another form of statigraphic trap. Pinch
out of beds against the flank of salt domes would
be a form of structural trap. Figure 3. Stratigraphic traps are formed in many ways, such as
buried beach sands, shoe string sands (sand bars
of former rivers), changes of facies and so on.

Until very recently, geophysical measurements located structures capable of trapping
petroleum, but drilling alone could prove their
presence or absence. In the last year or so,
seismic methods which sometimes detect the presence of shallow gas have been recognized. These
are amplitude and phase changes in reflected
sound waves brought about by the large difference
of sound velocity between gas filled and liquid
filled sediments. The economic potential of a
field must still be evaluated by drilling.

Regional drilling for stratigraphic information
is important in interpreting geophysical data.
Such drilling determines the types and ages of sediments in the region, whether or not source beds
are present, the possible existence of permeable

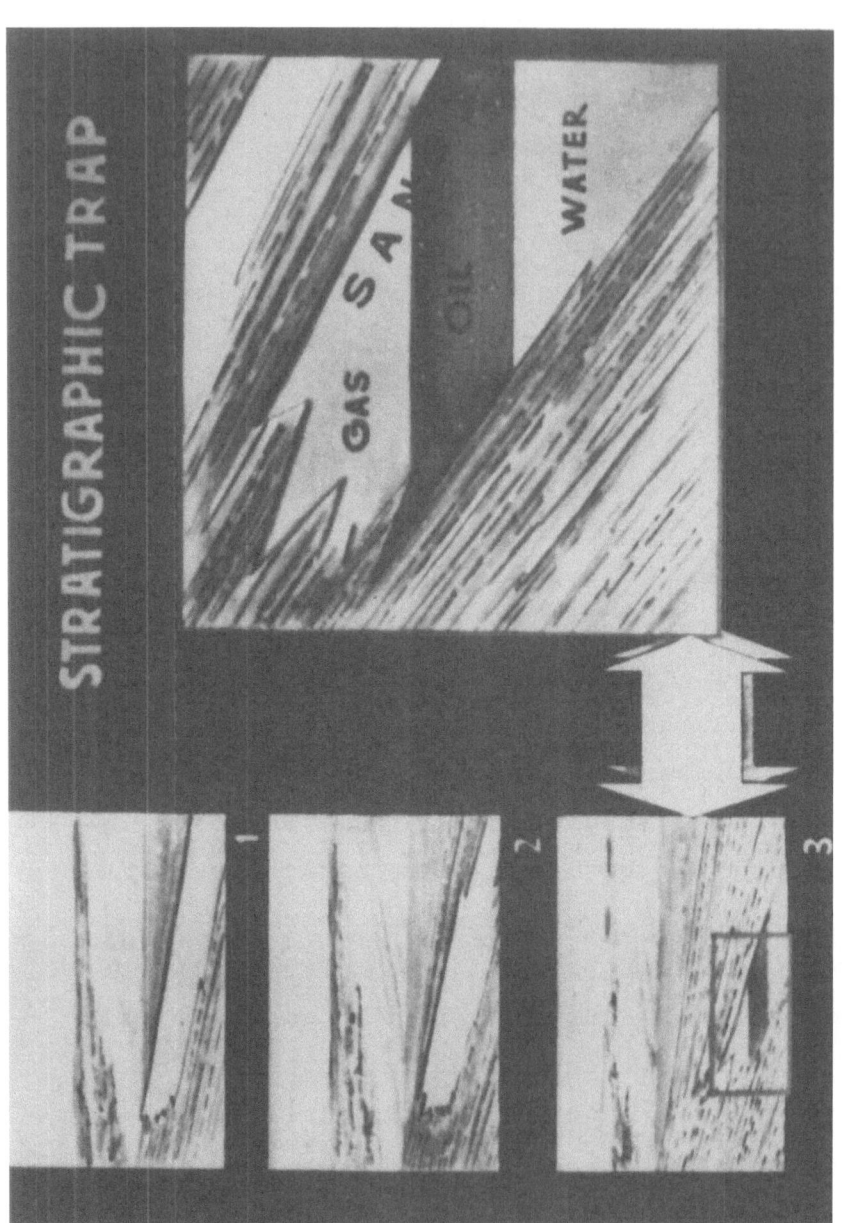

Figure 3.   Diagrammatic stratigraphic trap evolved from the burial of beach sand.   (After API)

beds for reservoirs the existence of impermeable beds so reservoirs could be formed, and occasionally gas and oil shows.

Offshore from the continents there are considerable bodies of sediments. Figure 4 shows one cross section through the edge of a continent into a deep sea basin. Usually there is a continental shelf which may be over 150 miles wide and whose surface is inclined seaward by slopes of the order of 0.1°. At depths of 150 meters to 300 meters, an increase of slope occurs to that of the continental slope. Where much sediment is present, as for instance off our east coast, this slope will be of the order of 3° to 6°. At the foot of the slope there is a decrease of slope to the order of 0.5°, known as the continental rise. In the ocean basins, seaward of the continental rise, the beds are normally very flat-lying with slopes of 0 to .01°. Figure 5 shows the world distribution of sedimentary layers on land and beneath continental shelves, cross-hatched and beneath the continental rises and slopes shown in the dotted pattern. Note the areas of the two patterns are about equal. Usually the sediments are thicker beneath the continental rise and slope so that there is more sediment volume in the deep offshore sector.

Sedimentary deposits in lakes and the seas are made up of the "rain of death" i.e. the residue of living organisms within the water column and on

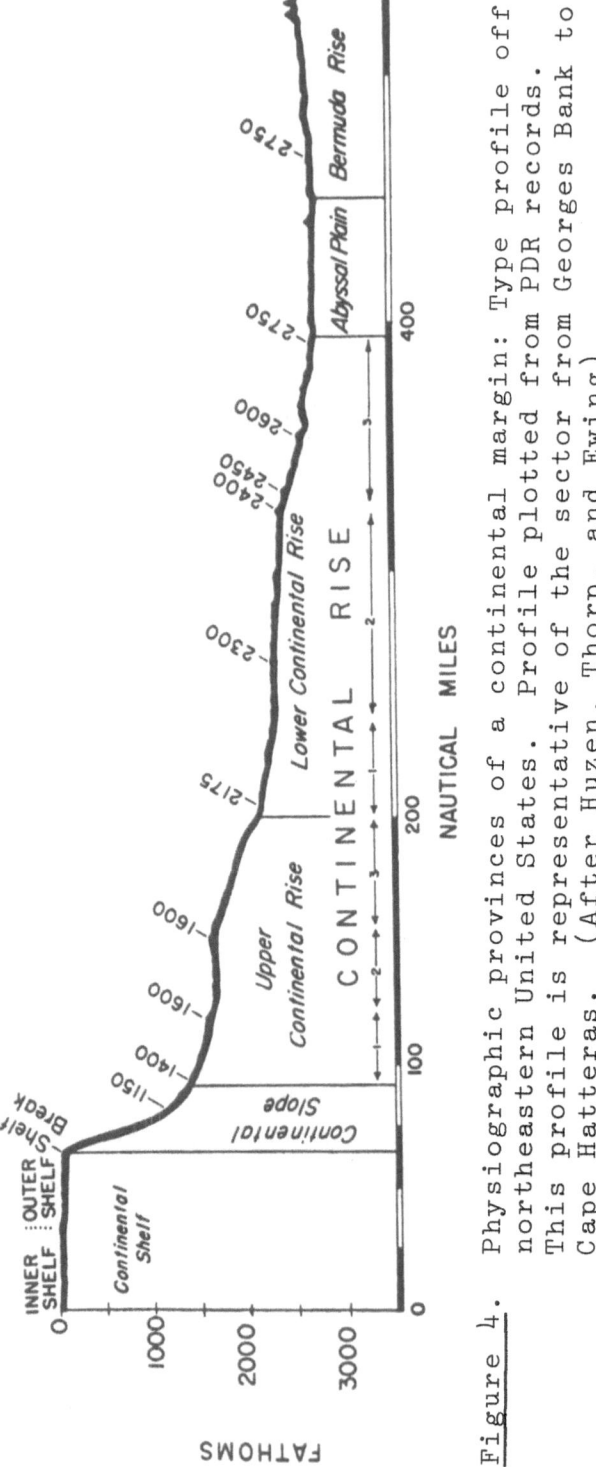

<u>Figure 4.</u>   Physiographic provinces of a continental margin: Type profile off
northeastern United States. Profile plotted from PDR records.
This profile is representative of the sector from Georges Bank to
Cape Hatteras. (After Huzen, Thorp, and Ewing)

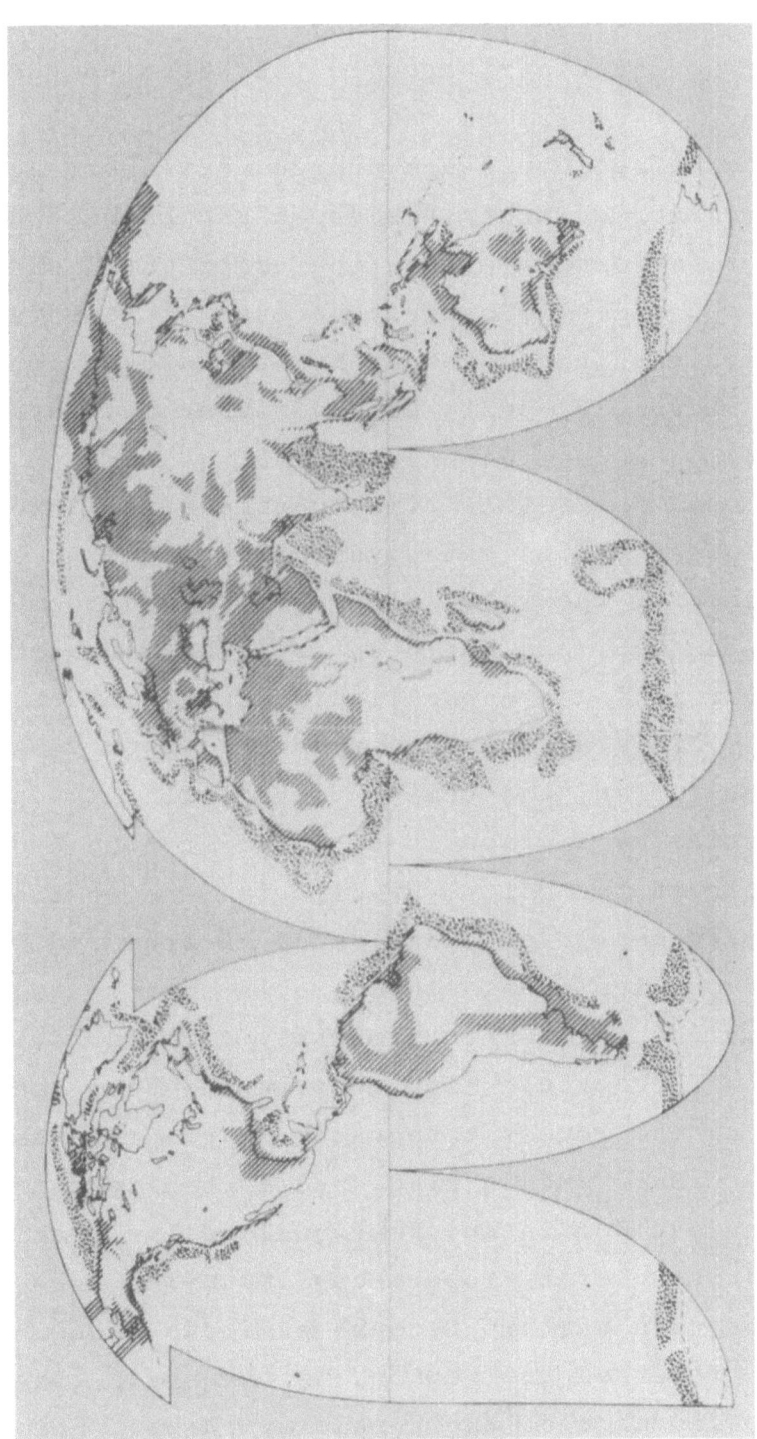

Figure 5. Areal distribution of thick sedimentary accumulation of the world. (After Emery)

bottom and the sediments eroded off the land delivered to and distributed in the water covered
area, mainly by rivers and wave action. Most
river borne sediments are dumped near shore and
then transported along the shore and seaward by
wave action forming at sea the continental shelves,
slopes and rises. Possibly because of de-watering
processes within the sediments, by wave action in
severe storms, by earthquake disturbances, and by
sea level changes, sediments near the edge of the
continental shelf from time to time slide into the
deeper ocean basins cataclysmically. Due to the
turbulence of the water near such a slide some or
all of the sediment would enter into suspension in
the water making a dense liquid which would flow
down the slope as a turbidity current, traversing
the slope, rise, and coming to rest finally in the
deep basins as a graded bed, i.e. with the coarse
sediments at the bottom grading to fine at the top.
At lowered sea level stand, such as occurred in the
Pleistocene ice ages, storms could stir up sediments on the shelves to similarly form turbidity
currents which would flow down to the deep basin
following the lowest topography. An earthquake
in 1929 caused such a slide and turbidity current
between Nova Scotia and Newfoundland. A section
of the continental slope about 100 miles long, slid
into the deep basin. Because numerous telegraph
cables traversed the region, to the south, the
speed of advance of the turbidity current across
the continental rise was evaluated as in excess of

100 miles per hour. A turbidite layer of sand three feet thick, which could be identified with that turbidity current, was deposited 300 miles seaward of the base of the slope, and the fringes of the turbidity current were traced more than 1000 miles south. When the telegraph cables were repaired, the broken ends were found to be splayed out for about 100 feet, the insulation and copper wires missing, and the steel wires highly abraded and polished. Sections of cable about 100 miles long in the central area were never found, presumably because they were transported too far away and/or buried too deep. Abundant evidence of such flows have been found in all sectors of the deep ocean basins and sand and coarse gravel deposits of several decimeters thickness have even been found in the deep ocean basins - no doubt transported there as turbidity currents. Drilling in the deep sea has discovered such layers in sediments of all ages so far penetrated (back to about 200 million years ago). Geologists on land have now identified layers with the same characteristics at least as far back as Cambrian times (500 million years ago). Some turbidity currents are also generated directly from river borne sediments. Turbidity current deposits are very prevalent in interior basins such as the Gulf of Mexico, the Bering Sea and the Mediterranean Sea.

Landslides and turbidity currents bring organic matter from land and the continental shelves into the deep sea environment and bury organic

matter living on the ocean floor. Green leaves
have sometimes been recovered from turbidity current deposits indicating their recent activity.
Figure 6 shows some of the living organisms on
the ocean floor. Ocean bottom photography shows
marine life or evidences of marine life in nearly
all of the one million or more pictures taken of
the deep ocean floor all over the world. Much
of this organic matter buried under the deposits
of these episodic events is protected from oxi-
dation by the rapid burial to greater depths than
ocean bottom feeders reach (a few centimeters).

Conceivably many more hydrocarbons are present in the deep sea than on land and beneath the
continental shelves, and that petroleum generation
is a continuous process still going on. Whether
sufficient quantities are accumulated in sealed
off reservoirs to make them commercially viable
remains to be seen.

In 1972, as reported by the Oil and Gas
Journal, oil production in the U.S. was 3.45 billion barrels, 86% from land areas and 14% from
continental shelf areas. Production on the continental shelves has slowly moved offshore until
some of the latest rigs have been located almost
out to the 200 meter curve. Figure 7. Some drilling rigs are bottom supported structures such as
the jack-up rig on the right, others are floating
rigs such as the semi-submersible rig in the fore-
ground or the drilling ship in the background.
Floating rigs are anchored with 8 or more ten ton

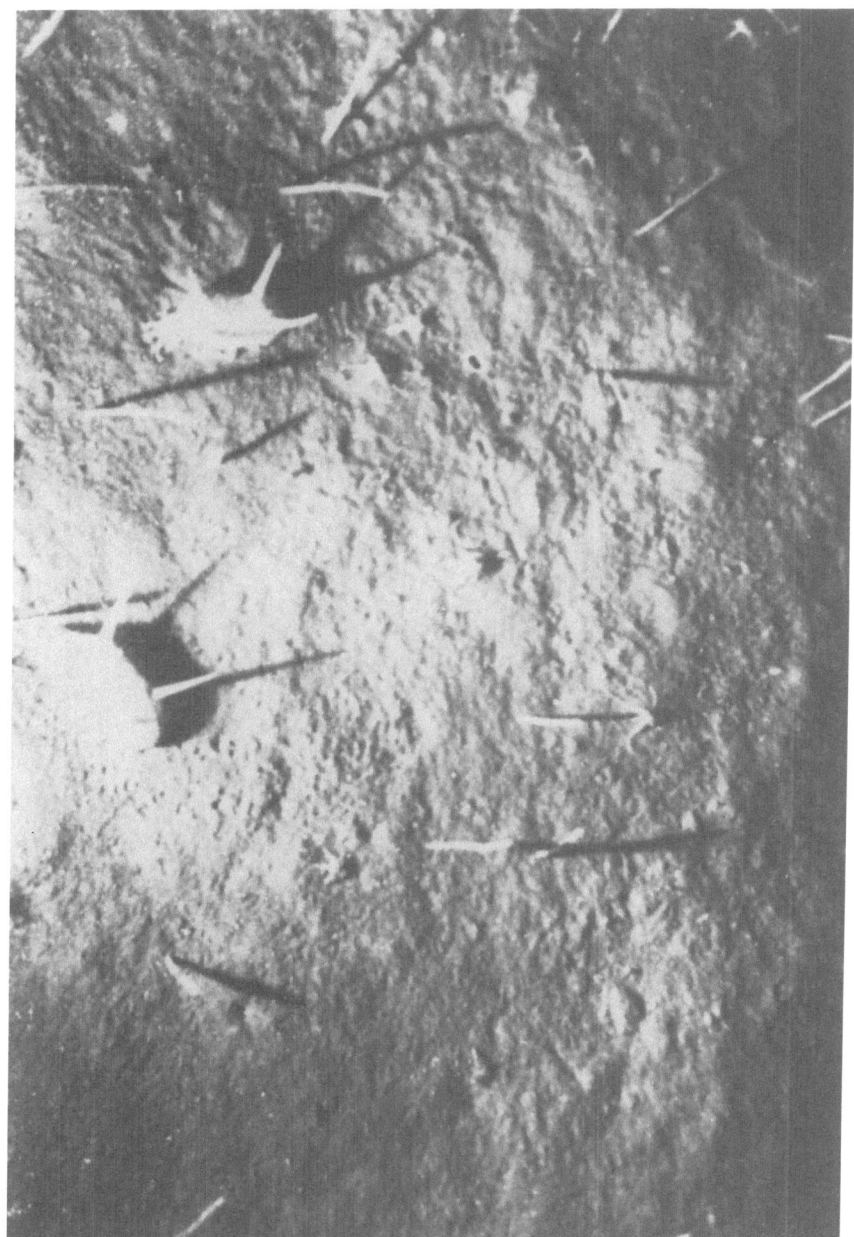

Figure 6. Ocean bottom photograph taken in 2800 fathoms of water depth.
Prevalent animal life is evident.

Figure 7.   Diagram showing jack-up, semi-submersible and floating drilling
rigs used for drilling for stratigraphic information and for
hydrocarbon production on the continental shelves.   (After API)

anchors. A single rig costs from 10 to 50 million
dollars. As many as thirty producing wells are
drilled from a single platform. Holes can be
routinely drilled to depths of at least 15,000
feet, and by angling holes they can be placed at
lateral distances of up to one mile from the platform.

Figure 8 shows a section from San Antonio,
Texas across the Gulf of Mexico to the Campeche
Scarp of Mexico. The section known from drilling
is indicated by the patterns on the left side.
The rest has been determined by seismic refraction
measurements. The numbers on the section are
sound velocities measured in km/sec. Numbers lower
than 5.5 km/sec certainly indicate sedimentary
layers, and in some cases, velocities below 6.5
km/sec may be sedimentary. The pinnacles repre-
sented in dashed lines are diagrammatic representations of regions where salt domes are present.
It can be seen that the volume of sediments is
greater beyond the 200 m (100 fm) curve than land-
ward of it. This is also true along the east
coast of the USA.

Drilling in deep water has been well established by GLOMAR CHALLENGER, Figure 9, which has
been drilling the ocean for scientific information
for about five years. A sonar beacon emitting a
ping once per second is placed on the bottom very
near to the desired drilling site. Three hydro-
phones located in a triangular arrangement on the
bottom of the hull receive these pings and a

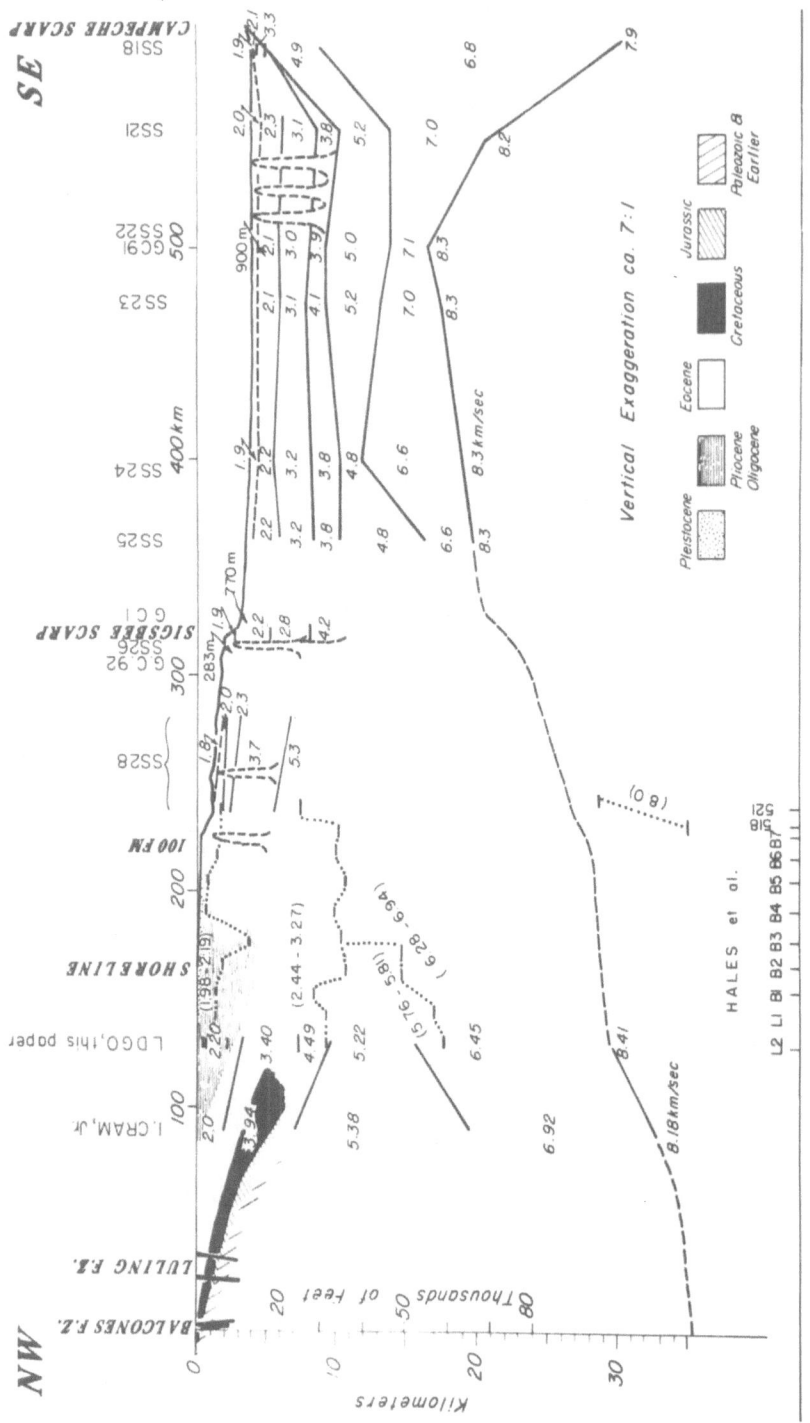

Figure 8. Structural geologic section through the crust southeast from San
Antonio, Texas to the Campeche Scarp off Mexico. (After Dorman,
Worzel, Leyden, Crook, and Halziemannuel)

Figure 9.   Deep sea drilling ship GLOMAR CHALLENGER used in the Deep Sea Drilling Project.

computer then resolves the position relative to the
marker and controls the main propulsion and bow
and stern thrusters to maintain the ship in any
chosen position (to an accuracy of about 50 feet)
within a circle ½ mile in radius from the epi-
central position. Reentry in the same hole, so that
drill bits can be changed, has been developed using
a large funnel about 7 meters in diameter across
the top with three sonar reflectors equally spaced
about the top of the funnel. A sonar transmitter
and receiver connected to the ship are placed in
the drill bit in place of the removable sampling
tube normally used. The return echoes are dis-
played on a cathode ray oscillograph. When the
ship is maneuvered into a position so that all
three echoes return at the same time, the drill is
stabbed into the hole, the funnel guiding the drill
bit laterally for the last few decimeters. Heave
compensation to correct for the vertical rise and
fall of the ship while drilling, has recently been
added to give better control of the drilling operations. When the hole is abandoned, it can be
plugged by back filling it with heavy mud or ce-
ment as the drill string is withdrawn the last
time.

   For production operations, blowout preventors
would have to be placed at the sea floor, tubes to
conduct the hydrocarbons to or near to the surface
or along the sea floor to shore would have to be
provided, and a "Christmas Tree", a complex valve

system, would have to be placed in the tube to control the flow of fluids. In some cases, pumps would also be required. Under some modes of operation, surface or near surface tanks for accumulation would also be used. Means for maintaining this equipment in position would also be required.

The Oil and Gas Journal has given estimated U.S. proven oil reserves of 36.4 billion barrels, 79% on land and 21% beneath the continental shelves. This would be a ten year reserve at 1972 production rates. Proven reserves are those which have been proven to be present by drilling and whose reservoir size is at least known by seismic means. Potential reserves are often quoted and represent "educated" and, perhaps in some cases, "uneducated" guesses. A simple volume of sediments type of estimate can be grossly misleading. There are no proven hydrocarbon reserves beneath the continental slopes, rises, or deep sea basins. So far as I know, no one has even estimated the potential reserves of these provinces.

Conceivably these could be many times more than the proven reserves above so it is important to quickly evaluate the deep sea sedimentary deposits. For instance, if the reserves per cubic mile in the slope and rise of the U.S. side of the Gulf of Mexico were the same per unit volume of sediment as beneath the shelves, there would be about 16 billion barrels of oil reserves present, or almost ½ as much as the total U.S. proven

reserves. At present, there is a three to four
year period from the location of proven reserves
and its production. There is little time to waste.

The continental shelves are essentially a
continuation of the continents beneath the fringes
of the sea. When we started our programs of geophysical measurements at sea in 1935, we showed
that continental shelves contained large volumes
of sediments and suitable structures for oil accumulation. We could not get support from the
petroleum industry because they said they did not
have the technology to produce oil at sea, costs
would be so high that it would be uneconomical and
because of the legal problems of ownership.

In 1945, the Truman Proclamation, that mineral
resources out to the 100 fathom curve along our
shores were under the jurisdiction of the U.S.
Government, changed the picture radically and since
then production on the continental shelves has
crept seaward until now wells are being contemplated in water depths near 500 meters (about 250
fathoms).

We carried out reconnaissance work in the
North Sea in 1953 showing that there were extensive sedimentary deposits there. Starting about
1962, drilling in the North Sea has been successful despite the difficult sea conditions.

We did the earliest Gulf Coast deep sea geophysical reconnaissance in 1953. We found large
volumes of sediments clear across the Gulf to the
Campeche Escarpment. A high velocity return from

shallow depth was observed beneath the continental
slope. We suggested that these could be salt or
granite intrusive features, but that we preferred
the salt dome interpretation. At that time, most
of the geological community would not accept the
salt interpretation. In the intervening years, it
has been demonstrated by many investigators that
the salt interpretation is correct and this is now
generally accepted.

In 1961 we introduced single channel seismic
reflection profiling in the deep water areas. We
discovered in that first work a region of dome-
like structures near the middle of the Gulf of
Mexico in water depths of 4000 m, Figure 10. On
the basis of seismic gravity and magnetic evidence we interpreted these as salt domes and suggested that petroleum would be present.

Once again the majority of the geological community refused to accept the salt dome interpretation and stated that oil could not be present in
the deep sea.

The second hole drilled from GLOMAR CHALLENGER
in the Deep Sea Drilling Project was located on one
of these "domes". Dr. Ewing and I were Co-Chief
Scientists for the first leg of the GLOMAR CHAL-
LENGER. Just before joining the ship, geologists
from one of the major oil companies showed me data
that they were convinced proved conclusively that:

1. Salt domes could not form in deep water
    areas.

2. If they were domes, they could not have

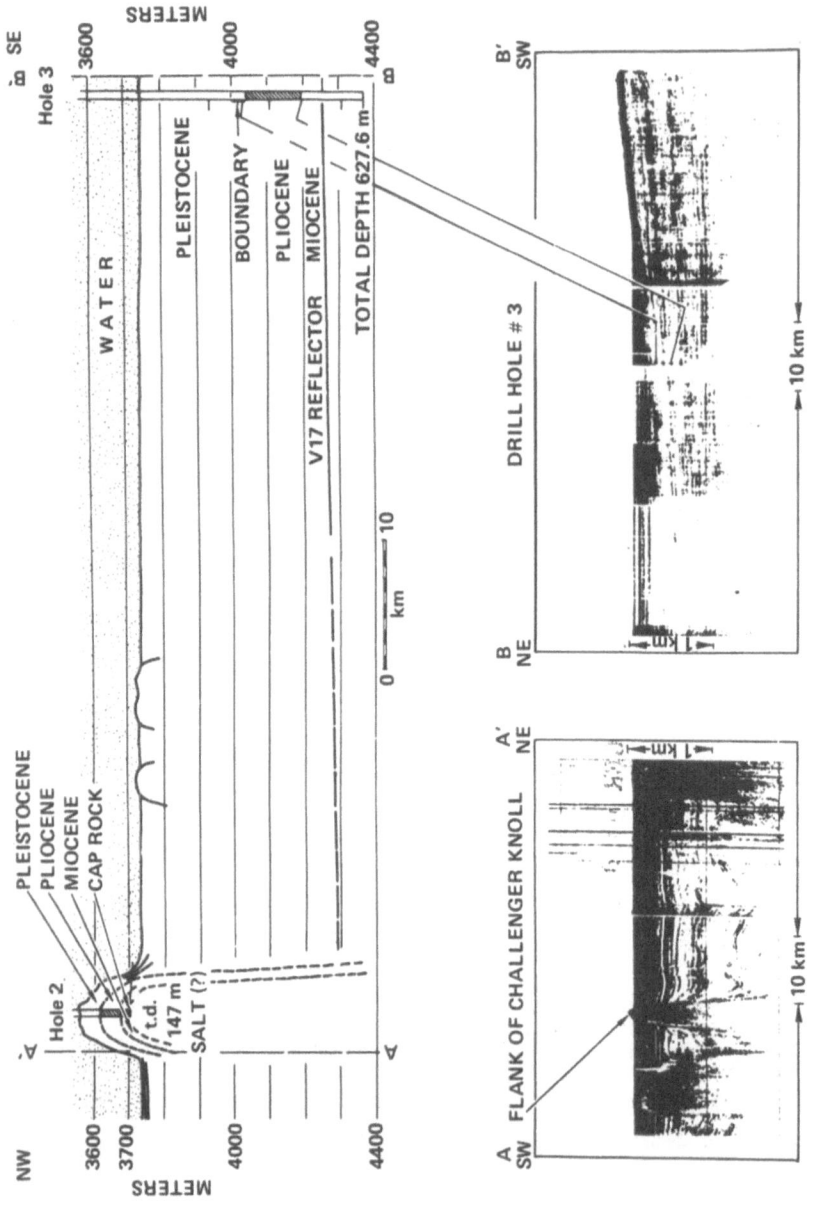

Figure 10.   Composit illustration made up of a diagram and reflection pro-
filer records of the central portion of the Gulf of Mexico.
(After Ewing, Worzel, and Burk)

cap rocks as cap rock was formed by perco-
lating ground water.

3. No oil could be present as it was formed
in the near shore environment.

Figure 11 shows how wrong they were. Dr.
Ewing and I are holding a core of cap rock from a
salt dome containing cap rock, sulfur, and liquid
petroleum. This core was recovered from 140 meters
below the ocean floor in 3572 m of water just 24
hours after my visit with the oil company geologists! It is only fair to state that other oil
company geologists (even in that same company) believed as we did that we would find salt domes,
cap rock and hydrocarbons.

This proved for the first time that oil could
exist in the deep sea areas.

Shortly after this, environmental concern
caused the formation of a "safety committee" for
the GLOMAR CHALLENGER drilling program, whose responsibility was to evaluate the possibility that
any hydrocarbons would be released into the ocean
from each proposed drilling site. In fact, holes
black-balled by that committee are good candidates
for any program to find petroleum in the deep sea.

Now it is generally accepted that petroleum
can be found in deep sea areas, but there is no
basis for estimating possible reserves - for a
lack of information.

Producing companies say (a) they lack technological capability to produce oil in deep water

Figure 11.   Scientific party of Leg I of GLOMAR CHALLENGER. Dr. Ewing and
Dr. Worzel, Co-Chief Scientists, are holding the core containing
cap rock, sulfer, and petroleum recovered from hole 2 on the salt
dome shown in Figure 10. This dome has now been named "Challenger Knoll" by the scientific party.

areas (b) costs of production would be formidable
and (c) a legal basis is needed to protect their
investment.

These are the same arguments used in 1935
against investigating the continental shelves.

PROPOSED PROGRAM

1.   We believe the petroleum industry has
done a superb job in producing oil to date by
gradually extending their operations across the
continental shelf.

2.   To stimulate them to make a big jump and
to evaluate and develop the deep water potential
there is needed

(a) Initiation and funding of a program of
regional geophysical reconnaissance of the
continental slopes and rises of the U.S.,
and the Gulf of Mexico and Bering Sea basins
to greater depths than single channel seismic
equipment has achieved. Multi-channel digital techniques developed by the exploration
companies principally for combating the multiple reflection problem on the continental
shelves has been shown to achieve the desired
penetration. Figure 12. We have just installed this type of equipment on IDA GREEN,
our research vessel, and the first non-industry
vessel with such equipment, to investigate
the transition zone between continents and
oceans.

Figure 12.   Research Vessel IDA GREEN showing the large reel on the stern
required for the detector cable for multi-channel digital re-
cording equipment now in use. This is the first non-industry
ship so equipped.

(b) Initiation and funding of a deep sea drilling program for stratigraphic reconnaissance, i.e. drill and test holes but if petroleum is present plug them now for later production. Proof of extensive reserves might well change the attitudes and prices of producing countries, and our need to depend on foreign sources in the future.

(c) Initiation and funding of a program to develop completion techniques for deep water areas. An agency independent of the government and the oil companies could expect to receive complete cooperation in personnel, equipment and know-how. The developed techniques should be made available to all.

(d) Economic incentives for drilling in deep water will have to be made attractive. This has partially been achieved by the OPEC nations doubling the price of crude oil with further price increases scheduled for the near future. More modest prices for leaseholds (a bid of over 200 million dollars was made recently for one lease off Florida) resulting in cheaper production prices would be achieved if larger leasing areas were made available. Small leases and small areas for lease artificially raise the competition for those leases available. Alternatively, a method of assigning leaseholds other than by bidding would reduce the costs. Finally, subsidies might be necessary.

(e) Legal support of the type of the Truman
Proclamation of 1945 to assure non-intervention
of foreign powers.

(f) Legal means for controlling nearby areas
of those that have been shown to be pro-
ductive to assure reasonable protection for
the investment of a producing well.

(g) Legal means so that anti-trust legisla-
tion does not prevent sharing the very large
risks of drilling in the deep sea by groups
of companies.

While no one can know the costs of these programs, the following ballpark estimates are of-
fered.

1. A one-ship regional program of geophysical
reconnaissance would cost between $3-5,000,000 per
year. Such a program is already started although
it is not funded for continuous operation.

2. A reconnaissance drilling operation would
cost about $10-20,000,000 per year for a single
drilling vessel producing about 20,000 feet of
hole with stratigraphic information. This could
be started immediately if an existing vessel could
be diverted. More than a dozen existing ships are
believed to have the capability.

3. Development of production techniques in
deep water (about 4 km) could start immediately by
drawing on expertise and perhaps equipment existing in industry. Costs are estimated at $150,000,
000.

4. Costs of drilling and completing a well in deep water after the technology is developed would be expected to be $3-10,000,000 per well.

5. Time and cost estimates of the legal needs would have to be obtained elsewhere.

These estimates could be made more accurate, quickly, after these programs start producing results. These programs would shed considerable information on the important scientific and practical question of the continuous generation of petroleum in the deep sea, and perhaps in the long run to the artificial generation of petroleum.

Henry Hurwitz, Jr.

FISSION POWER

H. Hurwitz, Jr.

General Electric Company

Schenectady, New York 12301

INTRODUCTION

My purpose is to briefly describe the status
and prospects of fission power. I hope that my
remarks will help to provide background for an exchange of ideas in the course of our panel discussions. Despite occasional representations to
the contrary, the power industry welcomes interactions with the scientific community, and I would
be glad to convey to my industrial associates any
well-considered viewpoints that arise from our discussions.

The information in Table 1 characterizes the
magnitude of the task of fulfilling anticipated
U.S. electrical power needs for 2000 AD. Note
the the size of the annual power cost and, even
more, the required capital investment in generating
capacity, indicates that economic considerations
cannot be ignored in the selection of power plant

TABLE 1

ESTIMATED U.S. ANNUAL RATES (~2000 AD)

Electrical Energy Generation 10^{13} KWH   (6 Times 1972)

Cost   > $200 x 10^9 (@ > 20 MILLS/KWH)

Heat Input 10^{17} BTU   (0.1 Q)

Number of Plants   1000+ @ > 1000 + MWE/Plant
                               (200 Utilities)

Assuming 60% Nuclear:

U$_3$O$_8$ Consumption   @ 1% Fissioned   1.2 x 10^5 Tons

Uranium Ore   @ 800 Parts/Million   150 Megatons
                          (1/20 Coal,
                          1/100 Oil Shale)

Note: Total Free World Nuclear Electric Energy to Date ≈ 0.2 Year at
        U.S. 2000 AD Rate.

types.

Because of the long lead time required in the
construction of electric power plants and the long
operating life of such plants, the general com-
plexion of the electric power industry is determined several decades in advance. Thus the electrical needs of the year 2000 AD must be fulfilled
to a major extent by plants being ordered at the
present time.* There is at present no assurance
that adequate amounts of fossil fuel can be ac-
quired, transported and utilized in a manner that
simultaneously satisfies environmental regulations
and is acceptable in cost. Therefore, nuclear
power must be relied upon to provide a major portion of our 2000 AD electrical needs. From this
point of view, it would appear that the primary
national priority in the energy field should not
be to find alternatives to present types of nuclear
power plants, but rather should be to support the
nuclear industry in its objective of assuring that
the needed nuclear plants will operate reliably,
safely, and economically.

CONTEMPORARY REACTOR TYPES

Several different nuclear reactor types have

---

*Since it is much more difficult to accelerate
than to decelerate the rate of construction of
electrical power plants, an underestimate of
future power needs could have severe societal
consequences.

already been utilized in the generation of significant amounts of central station electrical power.
While the cumulative generation of nuclear electrical energy to date is small in terms of the
anticipated 2000 AD annual rate, the experience
with nuclear plants has well demonstrated that
nuclear power can successfully take over the function of supplying base load electrical power.

The fact that different types of reactors have
predominated in different parts of the world is a
reflection both of varying selection criteria and
of historic circumstances. Experience has shown
that it is very difficult to make accurate com-
parative evaluations of alternative reactor types
prior to obtaining extensive operating experience.
In addition, the long development time of nuclear
power systems introduces the possibility that
selection criteria will change significantly in
the course of the development. Therefore ultimate
success in the development of a reactor system
cannot be assured by elaborate optimization studies
at an early stage in the program, but requires instead a continuous well funded and high quality
engineering and scientific effort throughout the
entire development cycle.

Canada made the selection of heavy water
moderated natural uranium fueled reactors at an
early date. The high moderating power and low
neutron absorption of heavy water makes possible
a high conversion ratio, i.e., the generation of
approximately 0.8 atoms of Pu^{239} for each U^{235}

atom destroyed. This, together with the development of a practical method for continuous re-
fueling, allows the natural uranium fuel to be
carried to an exposure of over 10,000 megawatt
days per metric ton in a "once through" fuel cycle.
In the Canadian deuterium-uranium (CANDU) reactors
the heat transfer medium is heavy water passing
through horizontal pressure tubes in which the
uranium oxide fuel element bundles are located.
Most of the heavy water moderator is located in the
region outside the pressure tubes. (The reentrant
structure that contains the moderator is referred
to as a calandria.)

The avoidance of uranium enrichment costs in
CANDU reactors is roughly compensated by the added
cost of the heavy water (over $50 million per
1000 MWe). The need to perform the isotopic separation step prior to reactor startup rather than
during the life of the reactor adds to interest
charges, but this disadvantage is minimized by the
relatively low interest rate paid by Canadian
utilities. The comparatively low power density in
the CANDU reactors tends to increase capital costs,
and this has lead to the early adoption in Canada
of the practice of building several reactors at
the same site.

Because of the high fixed charge rate in the
U.S., the development of serious interest in heavy
water reactors appears unlikely unless there were
to be a significant reduction in the cost of heavy
water and a major increase in the cost of uranium.

In addition, heavy water reactors, like light water
reactors, would be severely penalized by proposed
U.S. environmental regulations prohibiting the
consumptive use of water and thereby enforcing the
use of dry cooling towers.

The power reactor program in England, has,
like that in Canada, been based on the use of
natural uranium. The moderator in the U.K. reactors is carbon, and the heat transfer medium is
carbon dioxide with a core exit temperature of
about $700^{\circ}F$. While the U.K. gas cooled reactors
have successfully generated large quantities of
electrical power, it has thus far not been possible
to reduce the costs to an attractive level. Ac-
cordingly, there is now a vigorous debate in the
U.K. concerning the desirability of shifting to a
different reactor type, e.g., a U.S. light water
reactor, to fulfill short term power needs. Research and development would in any case continue
in the U.K. on high temperature gas cooled reactors and liquid metal fast breeders.

The longer range prospects for gas cooled
reactors has been considerably improved by two
developments not incorporated in the U.K. power
reactors. The first is the substitution of helium
for carbon dioxide, and the second is the development of fuel elements consisting of submillimeter
coated particles of uranium oxide or carbide
embedded in a carbon matrix. These concepts have
been tested in the Dragon reactor in the U.K.,
the Peachbottom reactor in the U.S., and the AVR

pebble bed reactor in Juelich, West Germany. The
larger (330 MWe) Ft. St. Vrain high temperature
reactor (HTGR) in the U.S. is expected to commence
operation at power during the year 1974.

Even with helium as the coolant, the HTGR's
do not achieve the high power density of light
water reactors. (The core power density in gas
cooled reactors is typically 5 to 10 kilowatts per
liter as compared to 50 to 100 kilowatts per liter
in light water (LWR) reactors.) To overcome this
difficulty, HTGR's have adopted the prestressed
concrete reactor vessel concept (PCRV). In this
concept, high pressure capability is provided by
steel cable tendons that keep the concrete under
compression. The reactor vessels for HTGR's are
typically designed to have a 33 foot internal
diameter and to operate at a pressure of 700 psi.
(This compares to the approximately 20 foot internal diameter and 1000 to 2000 psi operating
pressure of the steel vessels used for light water
reactors.) While the PCRV's help to eliminate the
onus of low power density, they require time consuming field construction, and effective thermal
insulation of the concrete from the high temperature helium.

The two major types of HTGR's are the prismatic and pebble bed designs. In the former
(marketed by General Atomic) the fuel is contained in large (14" x 31") hexagonal graphite
prisms provided with small cylindrical channels
for the passage of helium. These prisms are

arranged in a uniform array, and loaded and unloaded by special remotely controlled manipulators.
In the pebble bed design the coated fuel particles
are embedded in graphite balls approximately 2
inches in diameter. These pebbles can be continuously introduced into the core and unloaded by
gravity. The pebble bed concept is claimed to
allow higher temperature operation than the prismatic design because of less stringent dimensional
tolerances on the fuel elements. (The AVR reactor
has operated at $1700^{\circ}F$, as compared to typical
prismatic HTGR design temperatures of $1350^{\circ}F$.)
On the other hand, pebble bed designs typically
exhibit lower core power density than prismatic
designs so that they require a further increase in
reactor vessel size.

In the HTGR's the use of natural uranium is
precluded by the absence of a lattice structure
to reduce neutron absorption in the U^{238} resonances.
Uranium enriched to about 5% U^{235} can be used in
the coated particle fuel elements. An alternative
possibility is to use uranium at a still higher
level of enrichment, in combination with thorium.
(The thorium can be contained in separate coated
particles to simplify reprocessing.) In the latter
case, U^{233}, rather than plutonium, is produced by
neutron absorption in the fertile material. The
implications of this with respect to the fuel
cycle will be discussed later.

Despite the absence of experience in the construction and operation of full size ($\sim$ 1000 MWe)

HTGR's, several orders for such reactors have
already been placed by U.S. utilities. Selling
points for the HTGR as compared to light water reactors have been an anticipated improvement in
plant thermal efficiency (37-38% versus 33-34% for
light water reactors) and the possibility of somewhat better uranium utilization because of the
relatively low neutron absorption cross section of
carbon. These advantages would not be sufficient
in themselves to compensate for a substantial increase in plant costs, and HTGR's are thus being
sold at prices comparable to light water reactors.
It remains to be seen whether or not these prices
will be justified by actual experience in building
and licensing the plants.

Potential long range HTGR advantages include
the possibility of utilizing a direct cycle in
which the helium coolant is used as the working
fluid for helium turbines, and also the possibility
of developing economic process heat applications.
The direct cycle HTGR allows heat rejection to
take place at a higher temperature than with the
Rankine cycle LWR's and HTGR's, thus making the
use of dry cooling towers less costly.* On the
other hand, the direct (Brayton) cycle requires
a higher exit gas temperature from the reactor core
(above $1500^{\circ}F$), and thus further complicates fuel

*Note, however, that even with the direct cycle,
dry cooling results in a marked loss in plant
capacity at high ambient temperature.

element and structural material problems. In any
case, the direct cycle HTGR requires a lengthy
development program involving initial tests of the
turbine designs in fossil fired units, followed by
a series of demonstration and prototype plants.

The possible use of nuclear energy as a source
of process heat is receiving much attention throughout the world at the present time. Numerous candi-
date processes exist for using nuclear heat in the
chemical and metallurgical industries. By using
nuclear heat in the production of synthetic fuels,
the consumption of fossil reserves can be reduced
or eliminated entirely. In Germany a large pebble
bed reactor has been proposed for use in the re-
forming of methane. This process would convert
methane gas and water into a mixture of hydrogen
and carbon monoxide. The mixture could be trans-
mitted to a user who could catalytically recombine
it to produce methane, water, and heat. The
methane could be returned to the reactor site,
thereby providing a closed cycle for the distribution of energy. In Japan, nuclear heat is being
seriously considered for use in the steel industry. These nuclear heat applications appear
less imminent in the U.S. because fossil fuels are
not yet in such short supply as to justify the
added capital investment required to utilize
nuclear heat.

In the U.S. light water reactors dominate
the nuclear power industry. Because of neutron
absorption by hydrogen, it is necessary to enrich

the uranium fuel from 0.7% U^{235} to between 2% and
4% U^{235}. At the present time the enrichment is
carried out in diffusion plants that were developed in the course of the nuclear weapons program.
Pumping UF_6 gas through the diffusion barriers
requires considerable energy, so that a few percent of the electrical energy that will be derived
from the enriched uranium must be invested in
advance in the separation plant. While this re-
circulation of energy is small compared to unity,
there are economic incentives for reducing the
enrichment power requirements. This has occasioned
great interest in the alternative centrifugal and
laser isotope separation processes. In addition
to reducing energy requirements by at least an
order of magnitude, these latter processes require
fewer stages and are believed to offer the potential advantage of being economical in relatively
small size units. (By the same token, they are
feared by some as being a means of abetting nuclear
weapon proliferation.)

A key ingredient in the success of light
water reactors has been the development in the
U.S. Naval Reactor program of zirconium alloys
that can be used as core structural material and
fuel element cladding. In the commercial power
reactors, the fuel elements consist of short
cylindrical pellets of pressed uranium oxide that
are encased in zirconium tubes about 1/2" in
diameter. Light water reactors are of two
principal types which are differentiated by the

mechanism employed for steam generation. In the
boiling water type (BWR) the steam is generated
in the reactor core by boiling of the reactor
coolant. Low pressure jet pumps are utilized to
circulate the feed water through the core. A
small fraction of the water is vaporized during
each transit through the core, and after the
vapor has passed through steam separators within
the reactor vessel it is piped directly to the
steam turbine. In the pressurized water reactor
(PWR) the reactor pressure is about 2000 psi
rather than 1000 psi, so that the primary coolant
does not boil. External steam generators allow
heat to be transferred from the primary coolant
to water in a secondary system. The steam formed
by boiling of the secondary loop water provides
steam which serves as the working medium for the
turbine.

The BWR and PWR both produce low temperature
steam (550°F to 600°F) and thus require special
low pressure steam turbines. The PWR core power
density is over 100 kW per liter, which is about
twice that in a BWR. This leads to a smaller
(but higher pressure) reactor vessel in the PWR
case. There has been intense competition between
the BWR and PWR which has lead to rapid evo-
lutionary improvements in the designs. Since the
overall performance of the two LWR reactor types
is quite comparable, considerations such as plant
reliability and reputation of the manufacturer
are playing major roles in determining customer

preferences.

URANIUM UTILIZATION IN THERMAL NEUTRON REACTORS

The reactors now in commercial operation
make relatively little use of the fissile material
(i.e., Pu^{239} or U^{233}) that results from neutron
absorption in the fertile material (i.e., U^{238} or
thorium). Some of the newly generated fissile
material is consumed before the fuel elements are
unloaded. Recycling of the fuel by extracting the
fissile material remaining in spent fuel elements
and reloading it in fresh fuel elements is technologically feasible, but the required fuel reprocessing and refabrication plants are not yet in
operation. Although the recycling process is expensive, studies show considerable incentive for
carrying it out. In the case of light water reactors, recycling of fuel offers a 40% reduction
in uranium ore requirements and a similar saving
in isotopic separative work. Thus the anticipated increases in the cost of ore and separative
work will further increase the incentive for fuel
recycling. On the other hand, technological and
licensing problems, together with the difficulty
of acquiring capital for the needed facilities
tend to discourage the rapid introduction of recycling.

Recycling in principle provides greater
benefits in the thorium cycle since the U^{233} that
is generated from neutron absorption in thorium

has a higher intrinsic conversion ratio when used
in a thermal neutron reactor than Pu^{239}. Unfortunately, recycling of U^{233} is particularly difficult because of the U^{232} impurity that builds up
as a result of n-2n reactions. The U^{232} leads to
a chain of radioactive decays that release ener-
getic gamma rays. Hence refabrication of U^{233}
fuel must be performed in facilities that are remotely controlled and maintained. (This is more
difficult than the glove box procedures that are
required in the recycling of Pu^{239}.) It should
also be noted that while the U^{233} cycle uses
thorium rather than U^{238} as the fertile material,
it does not eliminate the fundamental dependence
of the nuclear fuel cycle on U^{235} that must be obtained from uranium.

A detailed discussion of thermal reactor fuel
cycles has been presented by Perry and Weinberg.[1]
These authors point out that thermal reactors
utilizing the U^{233} cycle can in principle make far
better use of uranium than is now being done in
the nuclear industry. Indeed there have been
numerous suggestions that thermal neutron reactors
utilizing U^{233} in conjunction with any of the
three most practical moderators (heavy water,
light water, or carbon) could actually approach
the breeding regime. Unfortunately, the design
and operating requirements for achieving this high
conversion ratio tend to be undesirable from the
economic standpoint. Low power density, frequent
reloading, and frequent reprocessing would be

required. Reactors designed for continuous re-
fueling, such as the CANDU reactor, the pebble bed
reactor, or, even better, the homogeneous fused
salt reactor with continuous reprocessing, would
be most suitable for achieving good fuel utilization. The lack of enthusiasm for pushing strenu-
ously toward better fuel utilization in thermal
neutron reactors results from unattractive engineering features of published designs for near-
breeder thermal neutron reactors. Furthermore, in
present light water reactors the ore cost con-
stitutes only about 5% of the nuclear power cost
so that even a small percentage increase in the
other costs could easily wipe out any saving due
to better utilization of ore. (The separative
work cost is also only about 5% of the nuclear
power cost, and this cost is not expected to
escalate by a large amount.) Although a fairly
rapid escalation in uranium ore cost is anticipated, it is argued that the fast reactor with its
far superior breeding ratio is a more reasonable
long-term solution than thermal neutron near-
breeder reactors. Indeed it has been pointed out
that fast breeders can initially be fueled by
fissile material produced in thermal reactors, and
could ultimately help to provide fuel for thermal
reactors if uranium ore becomes very expensive.
(Since plutonium produced in thermal reactors
would be more valuable as initial fuel for fast
breeders than as recycle fuel in thermal reactors,
the development of fast breeders would improve the

TABLE 2

<u>TYPICAL FACTORS INFLUENCING CONVERSION RATIO</u>
IN THORIUM-U^{233} CYCLE

$\eta - 1$ (U^{233})	1.24
Moderator Absorption	- 0.04
Fission Product Absorption	- 0.15
Leakage and Shim Control	- 0.10
Xenon Absorption and Override	- 0.04
Protactinium Absorption	- 0.04
Fuel Recycling Limitations	- 0.07
	0.80 (U^{233})

(NB - These loss estimates are intended to represent typical engineering values rather than
irreducible minima. Numerous tradeoffs involving
choice of reactor type, thermal efficiency, and
economics of construction and operation are
possible.)

economics of the thermal reactors by sustaining
the plutonium price level.)

Before leaving the subject of thermal reactors,
we shall give a semiquantitative discussion of the
potential fuel utilization of thermal reactors
utilizing the Th-U^{233} cycle. In a thermal neutron
spectrum the number of neutrons (η) emitted per
neutron absorbed in U^{233} is about 2.2. Since this
ratio is greater than two, a thermal neutron
breeder reactor is possible in principle. Unfortunately, there are several factors that de-
grade the conversion ratio in practical reactors,
as shown by Table 2. Using the 0.8 net conversion
ratio derived in the table, we may estimate the
utilization of uranium in a thermal reactor fuel
cycle as follows. Assuming that uranium ore is
fed into a separation plant with a waste stream
(tailing) concentration of 0.3% U^{235}, the U^{235}
actually recovered will constitute about 0.4% of
the initial uranium input. Assuming complete recycle in a thermal neutron spectrum, about 20%
of the U^{235} will capture a neutron to become U^{236},
while the remaining 80% will eventually fission.
For each U^{235} atom destroyed, approximately 0.6
atoms of U^{233} will be produced. (This is less
than the figure of 0.8 derived in Table 2 since η
for U^{235} is about 0.2 less than η for U^{233}.)
Each U^{233} atom will lead directly to about 0.9
fissions and also to 0.8 new atoms of U^{233}. The
total number of fissions as a percentage of the

original atoms of natural uranium is thus*
$0.4(0.8 + 0.6(0.9 + 0.9 \cdot 0.8 + ...)) = 1.4\%$.
With this percentage utilization of natural
uranium, about 80 tons of U_3O_8 would be required
to operate a 1000 megawatt reactor for one year.
(We have assumed an 0.8 capacity factor and a
10,000 BTU/kWh heat rate.) This uranium requirement is about one third that in a light water reactor with no recycle, and half that in a light
water reactor with recycle. It is also about 20%
less than the uranium input required for a CANDU
reactor with the "once through" fuel cycle.
(Note that since the CANDU reactor utilizes natural uranium it is not penalized by the loss of
U^{235} in the tailings of the separation plant.)
Some additional improvement in uranium utilization
would be possible by reducing the U^{235} concentration in the tailings of the separation plant. This
is one of the incentives for developing improved
isotopic separation methods.

It would appear from the above crude analysis that improvements in uranium utilization of
much more than a factor two over light water reactors with plutonium recycle is not likely in the
U.S. Therefore it would appear prudent to analyze

---

*For CANDU reactors, which are claimed to have a
potential conversion ratio of 0.92 if operated in
the thorium-U^{233} cycle, the utilization of natural
uranium would be about 4%. This corresponds to a
natural uranium requirement of under 30 tons per
1000 megawatt reactor per year.[2]

uranium requirements for the nuclear industry
under the assumption that until fast breeders are
deployed the consumption of ore will be approximately that characteristic of light water reactors
with plutonium recycle.

The pessimism expressed above with regard to
the potential impact of thermal neutron reactors
with improved fuel utilization is by no means
shared by all. The case for thermal neutron near-
breeders rests on the belief that such reactors,
together with the required fuel recycling facilities, can be built and operated economically, and
also that the price of uranium will rise to a
level that would severely penalize current reactor types but that would not make breeders imperative. If these assumptions turn out to be
valid and if the fast breeder program encounters
major delays, then greater interest in thermal
neutron breeders or near-breeders would be justified.

URANIUM AVAILABILITY

There have been numerous studies by government agencies and private organizations of the
projected price trend of uranium.[3] These studies
are based on similar assumptions about the future
uranium requirements of the nuclear industry. By
the year 2000 AD the U.S. requirements are
assumed to be of order of magnitude 200,000 tons
per year. (This assumes little impact of the

fast breeder, and approximately a thousand operating reactors with fuel requirements similar to
present light water reactors.) The studies indicate that a few megatons of U_3O_8 would be available in conventional uranium deposits at concentrations not far below the 2000 part per
million level that is now being mined. Thus
uranium supplies appear adequate to accommodate
the industry without breeders at least into the
first part of the twenty-first century. Although
the price of uranium in constant dollars is expected to increase due to the need to mine poorer
grade ores, the penalty would not be intolerably
great since ore costs at present constitute only
a few percent of the nuclear power cost.

The situation further into the twenty-first
century becomes increasingly obscure because of
uncertainties in the timetable for deploying
breeder reactors and also because of a major gap
in our knowledge of the abundance of uranium at
concentrations between a few hundred and a
thousand parts per million. Something is known
about uranium deposits in shales, phosphates, and
granites at concentrations of below 100 parts per
million, but it is doubtful that such deposits
could be worked economically in the foreseeable
future. (At 50 parts per million the energy content of uranium ore is about equivalent to that of
coal if the 1% utilization characteristic of nonbreeder reactors is assumed.)

Some semi-empirical theories of uranium

availability have suggested that there should be
enough uranium in concentrations above a few
hundred parts per million to fulfill requirements
at the 200,000 ton per year rate for as much as a
century. Unfortunately, these theories are little
more than guesswork, and should not serve as the
basis for long-range planning unless they can be
confirmed by an extensive program of exploratory
geological drilling.

Two other possible sources of fissile material
are natural uranium from the sea and plutonium
produced by neutrons from high energy accelerators
or fusion reactors. Uranium occurs in seawater at
a concentration of about three parts per billion
(i.e., one pound in 50 million gallons). Methods
have been proposed and partly tested for extracting uranium from seawater, but the cost
estimates range up to as high as several hundred
dollars per pound. Ion exchange beds with greater
flow rates per unit area and longer assured life
than those appearing in the literature seem
essential for achieving acceptable economics.

It is not known whether or not fissile
material could be produced by accelerators (utilizing neutrons from spallation reactions) or
fusion reactors at a cost low enough to be interesting for nonbreeder reactors. Investigations
planned within the next decade may help to answer
this question.[2,4]

THE FAST BREEDER

The scientific feasibility of the fast
breeder was established by the nuclear physicists
and chemists of the Manhattan Project during
World War II. It rests on the fact that Pu^{239}
emits on the average approximately 2.9 neutrons
per fission, and has only a small radiative capture cross section for unmoderated fission spectrum neutrons. Thus for each plutonium fission
one neutron is available to maintain the chain reaction, another to be absorbed in U^{238} to replace
the atom that has fissioned (by the reaction
$U^{238} + n \xrightarrow{\beta^-} Np^{239} \xrightarrow{\beta^-} Pu^{239}$), and 0.9 neutrons remain
to generate additional fissile material. Not only
is the 0.9 ideal breeding gain less marginal than
the 0.2 value for breeding U^{233} in a thermal neutron
spectrum, but also there is less parasitic capture by structural materials and fission products
in a fast neutron spectrum than in a thermal neutron spectrum. U^{233} can also be bred in a fast
spectrum but the breeding gain is somewhat less
than in the case of Pu^{239}.

On the basis of the favorable physics of
breeding with fast neutrons, it was relatively
easy to conceive of engineering approaches to the
design of fast breeder reactors. Enrico Fermi was
thus able in 1945 to deliver a stimulating collo-
quium talk at Los Alamos in which he described in
considerable detail the sodium cooled reactor concept that is now known as the liquid metal fast

breeder (LMFBR). This concept increases the
efficiency of uranium utilization by two orders of
magnitude and also makes possible the attainment
of the high thermal efficiency characteristic of
the steam conditions in modern fossil fueled power
plants. With the blueprint of a permanent solution
to mankind's energy problem so clearly in hand,
Fermi shortly thereafter chose to leave the field
of reactor physics and return to the challenging
problems of fundamental particle research.

As is universally the case in the implementation of new concepts, fast breeder development has
encountered problems that were not initially anticipated. Nevertheless none of these problems have
fundamentally compromised the attractive prospects
that Fermi was able to describe almost thirty
years ago.

From the nuclear physics standpoint, the main
disappointment has been that a relatively small
lowering of the neutron spectrum results in an increase in the Pu^{239} cross section for radiative
capture, thereby significantly reducing the
breeding ratio. On the other hand, some lowering
of the neutron spectrum is highly advantageous
from the safety standpoint since it increases the
magnitude of the negative Doppler coefficient of
reactivity which results from neutron capture in
the U^{238} resonances. This negative Doppler coefficient has the beneficial effect of tending to
shut the reactor down when the fuel temperature
increases thereby broadening the U^{238} resonances.

An experimental sodium cooled fast reactor
(SEFOR), sponsored jointly by General Electric,
the U.S. AEC, a group of U.S. electric utilities,
and the West German government, has shown that a
strong negative Doppler coefficient can be obtained without reducing the breeding gain to an
unacceptably low level.

Other problems have been encountered in regard to the engineering and economics of the
LMFBR. Because of the cost of manufacturing and
reprocessing fast breeder fuel elements, it is
necessary to be able to operate the fuel elements
for long periods without replacement (hopefully to
about 100,000 megawatt days per ton of heavy isotopes). This exposure level is not attainable
with plutonium and uranium metal, but fuel elements
fabricated from mixed oxides of uranium and
plutonium encased in steel tubing appear more
satisfactory. (These fuel elements are similar
to those used in light water reactors at lower
exposures.) Carbides of uranium and plutonium
would be superior to oxides because of their
higher heat conductivity, but it now appears that
special coatings would be necessary to prevent
contact between the carbides and the fuel element
cladding to avoid carbonization of the cladding,
or the formation of low melting point eutectics.

Swelling of fuel element cladding and core
structural materials due to the high flux of fast
neutrons is a significant problem in fast breeders.
Fortunately, the extensive research in this area

is bearing fruit, and the possibility of controlling the swelling by use of suitable alloys
has been demonstrated.

Because of the post-shutdown heat generated
in fuel elements due to fission product decay, it
is imperative to avoid any possibility of inadvertently uncovering the reactor core. Therefore present LMFBR design concepts all include
double wall containment of the sodium. This is
achieved either by surrounding the entire primary
system by a single large vessel ("pot" concept)
or by surrounding all individual components of
the primary system by guard vessels ("loop" concept).

In the safety analyses it is necessary to
show that hypothetical accidents will not cause
rupture of the outer vessels. While hypothetical
nuclear excursions are not considered to be serious
threats to the system, there has been concern that
the melting of a fuel element would lead to unacceptably high pressure transients because of
vapor that results when the ceramic fuel element
cores become mixed with the sodium. Actual fuel
melting experiments in a test reactor sodium loop
exhibit transients far below the theoretical
maxima, and it is hoped that further experiments,
together with refinements of the theory, will
provide increased confidence that fuel-coolant
interactions do not jeopardize the integrity of
the primary coolant loop.

In order to minimize the chance of

interactions between the steam generator and the
primary system, a secondary sodium loop is provided between the primary system and the steam
generators. While the isolation resulting from
the intermediate heat exchanger prevents steam
generator leaks from affecting the core, such
leaks would cause serious operating problems.
Therefore it is of utmost importance to develop
steam generator manufacturing and testing techniques that can provide the required reliability
at acceptable cost. While this objective is
generally considered to be technologically achiev-
able, it may well turn out to be the pacing item
in commercial fast breeder development.

Several large LMFBR reactors have been built
throughout the world. The largest of these are
the 250 MWe Phoenix reactor in France which
commenced operation in December 1973, the BN 350
reactor which has operated for several years in
Russia, and the Prototype Fast Reactor (PFR) in
England which should begin operation in 1974.
(The BN 350 reactor has recently been reported to
be out of service because of a serious steam
generator failure.) The U.S. Clinch River demonstration fast breeder is several years behind its
European counterparts, but it is hoped that the
more advanced feature of this reactor, together
with the research and development associated with
the U.S. LMFBR program will maintain an adequate
U.S. position in the fast breeder field.

Despite the preponderant world effort on

sodium cooled fast breeders, there have been recent
suggestions that a gas cooled fast breeder (GCBR)
would be preferable in the long run. Unfortunately
the carbon fuel elements that appear attractive
for HTGR's are not applicable to fast breeders.
Hence, it is proposed to use fuel elements similar
to those in the LMFBR. This forecloses the
possibility of reaching temperatures required for
direct cycle operation or process heat applications
which, as we have pointed out, are the major
potential advantages of thermal neutron gas cooled
reactors. Nevertheless, proponents of the GCBR
cite the improved breeding ratio that results from
the absence of sodium, and claim that despite the
low fuel element heat capacity and limited heat
transfer capability of the helium, hypothetical
accidents involving core depressurization or
failure of the helium circulators can be safely
accommodated. Extensive tests would be necessary
to verify this claim.

From the above discussion it is clear that
despite the several decades of fast breeder development effort throughout the world, the remaining
tasks in developing a commercially acceptable
fast breeder are difficult and time consuming.
The argument usually advanced for pushing ahead
energetically on the fast breeder program is that
the potential benefit in reduced uranium consumption resulting from early fast breeder deployment would amount to tens of billions of
dollars in the next few decades. A more compelling

argument is that there is at present no assured
practical alternative to the fast breeder for fulfilling energy needs beyond the early part of the
21st century. Therefore since the deployment of
fast breeder reactors will probably require a
substantially longer period of time than most
current estimates indicate, it is of utmost importance to push ahead as rapidly as possible. The
loss in momentum that would result from a deli-
berate slow down in the program would be difficult to reverse if the impracticality of long-
range alternatives to the fast breeder were to
become more obvious.

NUCLEAR SAFETY

In this brief discussion it is not feasible
to add significantly to the enormous literature
that has grown up on the topic of nuclear safety.
Instead, we shall restrict ourselves to men-
tioning the key issues and adding a few general
comments. The three main concerns that have been
expressed with regard to the safety of fission
power are possible harmful effects of normal low
level radioactive effluents from nuclear plants,
the conceivable possibility of catastrophic release of radioactivity, and the problem of nuclear
waste management. More recently concern has been
expressed about the possible diversion of fissile
or radioactive material from commercial plants
for terrorist purposes. We shall consider each of

these items in turn.

The fundamental point with regard to normal low level radiation release from commercial nuclear installations is that the resulting radiation exposure to the public is not only small compared to the natural radiation background but also small compared to fluctuations in the natural background. Despite this fact, it has become popular to perform numerical calculations of added injury to the public from nuclear power using data on radiation doses orders of magnitude larger than those under consideration, together with the assumption of zero threshold for injury by radiation. This leads to the conjecture that substantial injury is being caused to the public by natural radiation and that a small but finite increase to this background rate would result from large-scale deployment of fission power. Despite the fact that direct measurement of the effect of background radiation on the public is difficult, if not impossible, it is sometimes argued that all potential hazards that have neither been proved nor disproved should be avoided. Unfortunately, there is a denumerably infinite number of such hazards, and there is no clear-cut means of singling out the finite subset that should really concern us. In any case, there have been recent attempts to estimate injury to the public from either foregoing adequate power or from utilizing alternative power sources to nuclear energy. It seems relatively easy to come up with larger amounts of

estimated public injury from either of these
latter options,[5] and such calculations will no
doubt be featured prominently in environmental
impact statements for proposed fission power
plants.

The need to protect the public from large-
scale release of radioactivity from a hypothetical
accident to a fission power plant was recognized
at the time of inception of the nuclear industry.
It soon became apparent that relying for protection on distance alone was an untenable approach.
The first fission reactor of appreciable power to
be located outside a government reservation was
the SIR reactor which was constructed in the late
forties at West Milton, N.Y. Because this location was not as isolated as previous reactor sites
such as Hanford, Wash., it was judged prudent to
surround the reactor with a large steel Horton
sphere. Once this decision was reached, it became
necessary to establish engineering specifications
for the containment. This lead to the consideration of a number of hypothetical accidents to the
plant. It was possible to show that the energy
release of plausible nuclear excursions would be
smaller than the energy release from chemical reactions that could conceivably be initiated by
various hypothetical accidents to the plant. Hence
the containment specifications could be established
fairly quantitatively on the basis of the material,
chemical, and thermal properties of the reactor
constituents.

This general situation has prevailed in all
the subsequently constructed power plants. In
light water reactors the containment specifications are based primarily on the flashing of steam
from a hypothetical double ended rupture of the
largest pipe in the primary system. Catastrophic
rupture of the reactor vessel itself is not used
in establishing containment specifications on the
basis that rapid rupture of the reactor vessel is
excluded by the metallurgical properties of the
vessel material and that systematic inspections
of the vessel during construction and throughout
its life would show up major flaws long before
leaks of significant magnitude would be initiated.

In early commercial reactors containment was
provided by a single large steel vessel. (This
approach is known as dry containment.) It was
subsequently recognized that, particularly in the
case of BWR's, the cost of the containment could
be greatly reduced by a double structure consisting of a high strength but comparatively small
inner vessel (known as the dry well) and an outer
structure designed for lower pressure. Steam released into the dry well is vented through a
"pressure suppression" pool of water contained in
a toroidal region surrounding the dry well.
Large-scale tests have shown that the steam is
efficiently condensed by being directed into the
pressure suppression pool. In addition to reducing the volume of high pressure steam that
must be accommodated, the pressure suppression

system provides improved protection against
missiles by virtue of the massive high strength
dry well which surrounds the reactor vessel and
the critical pumps, pipes and valves. (In recent BWR designs the dry well is constructed of
reinforced concrete, and secondary containment is
provided by a 3 foot thick cylindrical, 136 foot
diameter reinforced concrete structure surrounding
a steel shell designed for 15 psig.) Alternative
schemes are available for condensing the steam
released by a hypothetical pipe rupture. For example, some PWR plants use the latent heat of
melting of ice to absorb energy from the steam.

The containment concepts described above
constitute the third level of defense against
public injury from nuclear accidents.[6] The first
level is provided by sound plant design, construction and operation. (The strong negative
temperature coefficients of reactivity characteristic of LWR's is in this category.) The second
level of defense is provided by engineered safeguards that prevent or minimize harm to the plant
and its operators in the event of unanticipated
failures in the first level of defense.

The emergency core cooling systems (ECCS)
provided in all LWR designs are prime examples of
engineered safeguards. The fundamental purpose
of ECCS is to prevent damage to the fuel elements
in the event of loss of coolant due to a hypothetical rupture in the primary system. While a
complete core meltdown following a loss of coolant

accident could conceivably jeopardize containment,
there is little likelihood that a partial core
meltdown would have significant off-site consequences. Nevertheless the conservative philosophy
has been adopted of demanding not only that the
ECCS be capable of preventing any fuel melting at
all, but also that large margins be maintained
between peak fuel cladding temperatures reached in
a hypothetical loss of coolant accident and the
temperature at which fuel element failure would
be expected to occur. (This conservation is
designed to eliminate any possibility of a steam
explosion due to rapid mixing of the oxide fuel and
water.) In order to assure that the ECCS would
reliably achieve the specified performance, redundant independent systems are provided for in-
jecting additional coolant into the core. Successful large-scale experimental tests (including tests
with full-size fuel bundles operating at and
above normal reactor power densities) have been
made of ECCS systems by reactor manufacturers and
the AEC. These tests have shown that the present
ECCS designs are capable of providing the desired
core protection under the postulated circumstances.

     In order to assure that the three levels of
defense cited above will be effective in a particular nuclear power plant, elaborate licensing
and plant inspection procedures have been in-
stituted by the government. While the licensing
procedures undoubtedly serve as a reassuring
check on nuclear plant safety, there has been

widespread criticism that these procedures have
become altogether too time consuming and unwieldy.
It has not been possible to avoid a trend toward
a legalistic adversary format that does not appear
effective for assuring that plant design is
optimum from the safety standpoint. Reactor design engineers are in many cases forced to expend
disproportionate effort in responding to specific
objections advanced by intervenors, and the
engineers become constrained in the use of their
own experience and resourcefulness to improve the
safety of their designs. Under the present
atmosphere, design changes, even when proposed for
the purpose of improving safety, can result in
delaying or reopening licensing procedures. Thus
the reactor vendors who are most interested in
achieving improved safety may in effect be
penalized rather than rewarded for their efforts.

The reactor vendors have constructed expensive facilities of their own for verifying the
safety of their product. For example, the 17.2
megawatt Atlas BWR test facility in San Jose is
the largest heat transfer loop in the world. Unfortunately, there is a tendency for favorable
confirmatory test data generated by the reactor
vendors to be given inadequate credence in
licensing procedures. Data presented by the manufacturer has not achieved the hoped for effect of
allowing some reduction in the redundant safety
factors imposed by the licensing procedure. Indeed, in some cases data provided by the

manufacturers has been used to further ratchet the
licensing requirements. While some improvements
in the speed and efficiency of nuclear plant
licensing should result from procedural changes,
the greatest improvement would result from a
demonstration on the part of intervenors that they
are objectively interested in improving the opera-
bility and safety of nuclear power plants rather
than in expressing their opposition to nuclear
power itself.

A longer range but equally controversial
issue in regard to the safety of nuclear power re-
lates to the methods of handling nuclear wastes.
Most members of the nuclear industry are firmly
convinced that this problem is solvable, and
furthermore that there will be adequate opportunity
during the coming decades to select and refine a
satisfactory approach to the problem. In some
countries (for example, Canada) there is no inten-
tion to do anything more than store spent fuel
elements until such time as economic considerations
justify the recovery of the plutonium that the
elements contain. Environmentalists, on the other
hand, have taken the fact that no ultimate solution to the waste disposal problem has been agreed
upon as indicating that no satisfactory solution
exists.

An aspect of the waste disposal problem that
is not adequately appreciated by the public (and
even some distinguished physicists) is that
fission products decay. In fact this natural

decay in most cases destroys the fission products
at a rate that probably exceeds the rate at which
these fission products could be burned in a hypothetical fusion reactor. Because of the natural
decay, the amount of radioactive waste that would
result from a stable nuclear economy does not grow
indefinitely. Instead, the amount of long lived
waste, measured in terms of the rate of radioactive energy release, approaches an asymptotic
upper limit equal to about 0.03 percent of the
fission power produced in the operating reactors.
This means that it should be relatively straight-
forward to store this waste in suitably engineered
and protected repositories associated with each
fission reactor complex. In order to avoid a continuously increasing volume of nuclear waste, it
would be possible to occasionally (e.g., once
every fifty years) reprocess the waste to re-
concentrate the remaining radioactive fission products. Fission products having exceptionally long
half lives (e.g., ^{99}Tc with a half life of 2 x 10^5
years) could be burned in operating nuclear reactors. The same is true of the actinides (trans-
uranic elements) that are produced by successive
neutron captures in the nuclear fuel. (The
actinides have been cited as being particularly
hazardous because of their long half lives, their
alpha activity, and their tendency to be con-
centrated in the skeletons of living organisms.)
Since some of the actinides are fissile isotopes,
it is actually beneficial to include them in fuel

elements. Loss of reactivity due to the non-
fissile actinides would be less in fast breeders
than in thermal neutron reactors.

Techniques for concentrating and solidify-
ing high level radioactive wastes have been well
developed, and are in the process of being
commercially deployed. The binding material for
solidified nuclear waste can be either a ceramic
(e.g., aluminum oxide) or a polymer (e.g., urea
formaldehyde). In either case a nonsoluble, non-
crumbling material is produced. This material can
be further protected by being encased in hermetic
steel cylinders. Depending on the initial concentration of the radioactive waste, the cylinders
may be stored in air or in a pool with some pro-
visions for cooling during the first few years.
In either case, continuous monitoring of the
cylinders would be provided to detect for unanticipated leaks.

Most people still feel that the concept of a
permanent repository for nuclear wastes is
superior economically, and perhaps also philo-
sophically, to permanent engineered storage.
Nevertheless, since the selection of a permanent
waste repository does not appear to be imminent,
the fuel reprocessing centers are required to
plan for storage of the high level wastes for an
extended period of time. The extrapolation to
permanent engineered storage would thus not be a
major one.

It is frequently implied that we still have

the option of avoiding the nuclear waste management problem by abandoning nuclear energy. This
is not really the case for two reasons. In the
first place, large amounts of nuclear waste have
already been produced by several nations in the
course of nuclear weapons and power programs.
While the existing amounts of nuclear waste are
still two or three orders of magnitude less than
the asymptotic quantity that would result from a
fully developed worldwide nuclear industry, the
qualitative nature of the waste management problem
would not be changed in a fundamental way. Indeed, the improvements in waste management techniques that would undoubtedly be stimulated by a
maturing nuclear industry could well result in an
actual decrease in the hazard to humanity from
nuclear wastes. A second point is that hazards
from nuclear wastes do not respect international
boundaries. Thus no nation has the power to unilaterally wish away the problem. Other nations
of the world have an even greater stake in the
success of nuclear power than the United States,
and they would not be likely to follow any U.S.
decision to slow down the development for nuclear
power.* On the other hand, if the United States
takes the lead in developing and utilizing a sound
nuclear waste management technology, then other
nations would be inclined to follow our example.

---

*The Chinese, for example, are reported to regard
U.S. concern over fission products as a capital-
istic phobia.

Therefore our safest course of action is to stop
bemoaning the difficulties of nuclear waste management and to embark on a vigorous research and
development program to bring the problem under
control.

Recent articles in the lay press have drama-
tized the possibility of diversion of fissile
material or nuclear waste for terrorist purposes.
While these articles are correct in their thesis
that utmost care must be exerted in preventing
such diversion, they considerably overstate the
ease with which nuclear materials could be used
for terrorist purposes. The slightly enriched
uranium oxide that fuels most of the commercial
reactors in the U.S. is not a suitable bomb material. Furthermore, the plutonium produced in the
nuclear industry is highly contaminated with Pu^{240}.
Neutrons produced in spontaneous fission of Pu^{240}
would cause a crude atomic bomb to predetonate so
badly that the nuclear energy release would be
unlikely to exceed that of the high explosives
used in attempting to set if off. Thorium cycle
reactors could initially be fueled with denatured
plutonium, and the U^{233} that they produce has
such a strong gamma activity that attempting to
fabricate it into a bomb would be a hazardous
undertaking.

Stolen plutonium or liquid nuclear waste
could conceivably be used to produce dangerous
local radioactive contamination. Therefore these
materials must be carefully guarded. If adequate

precautions are taken, the diversion of radioactive materials can be made so difficult and un-
palatable an undertaking that the danger to the
public would be virtually nonexistent.

   As in the case of nuclear waste management
problems, the diversion of fissile or radioactive
materials cannot be limited by international
boundaries. Therefore it is far better for the
U.S. to provide a good example to other nations by
maintaining strict control of dangerous nuclear
materials than to make a futile attempt to back
away from the problem by unilaterally relinquishing
the benefits of nuclear energy.

   The import of this discussion on nuclear
safety can best be summarized by repeating a
cogent observation that was recently made by
Samuel Goudsmit.[7] He pointed out that international
squabbles over the dwindling world fuel resources
might well trigger a nuclear war. Since the widespread deployment of nuclear power could help to
avert a major worldwide energy shortage, there
would be a corresponding reduction in the pressures leading to international conflicts. It is
clear that benefit to humanity of even a small
reduction in the probability of nuclear war would
far outweigh the minimal risk of harm to the public from the peaceful uses of nuclear energy.

REFERENCES

1. A.M. Perry and A.M. Weinberg, "Thermal Breeder
    Reactors," Annual Review of Nuclear Science,
    Vol. 22, p. 317 (1972).

2. J.S. Fraser, C.R.J. Hoffmann, and P.R.
    Tunnicliffe, "The Role of Electrically Pro-
    duced Neutrons and Nuclear Power Generation,"
    Atomic Energy of Canada Limited Report
    AECL-4658 (October 1973).

3. R.D. Nininger, J.A. Patterson, F.P. Baranowski,
    "Nuclear Fuel Resources and Requirements,"
    United States Atomic Energy Commission Report
    WASH-1243 (April 1973).

4. R.N. Horoshko, H. Hurwitz, and H. Zmora,
    "Application of Laser Fusion to the Production
    of Fissile Material," Annals of Nuclear Science
    and Engineering (in press).

5. R.P. Hammond, "Nuclear Power Risks, "American
    Scientist, 62, 155 (1974).

6. United States Atomic Energy Commission Report
    WASH-1250, "The Safety of Nuclear Power Re-
    actors (Light Water-Cooled) and Related
    Facilities," (July 1973); cf. also review of
    WASH-1250 by S.E. Rippon, Nuclear Engineering
    International, 19, No. 212, p. 25 (January
    1974).

7. S.A. Goudsmit (communication), Bull. Atomic
    Scientists, 30, No. 2, p. 2 (February 1974).

ENERGY BLACK HOLE

Behram Kursunoglu

Center for Theoretical Studies

University of Miami, Coral Gables, Fla.

ENERGY INDETERMINACY AND ECONOMICS

In the short and long run, the existence of everything in the universe is a function of energy. The time span of man's survival on an earth with finite resources will be determined by the ingenuity and wisdom which he exerts in using energy.

The economic aspects of the energy problem can be condensed into an equation of state where

$$Energy = Labor + Resources + Capital, \qquad (1)$$

and where each term on the right hand side can be measured in units of energy. For example, if we express capital in terms of gold, then its mining, manufacturing, and eventual minting into coins to be used as a monetary exchange can be measured in units of energy. Capital when it exists and is

135

used, brings about an interaction between labor
and resources. The energy resulting from the sum
of the three terms in the equation of state (1)
leads to the production of other forms of energy.
This energy can further be decomposed (approximately) into three parts

$$E = E_s + E_t + E_{st},\qquad\qquad(2)$$

where E_t is the energy lost (or disposed) into the
environment and is not directly recoverable such
as in the case of fossil fuel used in all kinds of
vehicular transportation and other devices which
consume fossil fuel to produce work. The term E_s
represents the gain in useful energy and is a
residual amount left over as a real asset to be
integrated into the survival process of man. The
amount of energy E_s is indeterminate.

  *All natural processes are subject to some
kind of generalized principle of invariant indeterminacy.* For example, it is an invariant that
either a boy or a girl is conceived by the mother,
but there is the indeterminacy as to whether a
boy or girl is born.

  The term E_{st} in equation (2) refers to the
energy of interaction between E_s and E_t and is
itself the cause of an indeterminacy in E_s. The
value of E_s is not defined by the actual in-
ventories of fuel resources alone but by resource
constraints based on geopolitical, environmental,
and above all, on the degree of participation of

all other people in energy consumption. The
energy consumption in one form or another is an
invariant requirement for survival but the amount
of usable energy is indeterminate. A further
reason for this indeterminacy is due to the rela-
tionship between the quality of the environment
and the pollutants which are emitted into it. For
example, electricity generating plants emit more
than half of the sulfur oxide produced by fossil-
fueled industries. Well-balanced electricity
generation from conventional plants requires that
discharges like carbon dioxide, oxides of nitrogen
and sulfur be minimized. However, the minimization of these emissions requires the establishment of some criteria for the atmospheric concentrations of sulfur oxides, hydrocarbons, carbon
monoxide, nitrogen oxides, photo-chemical oxidants,
etc. A faithful implementation of these measures
would lead to greater consumption of fossil fuel
per kilowatt-hour for the production of clean
energy and therefore to a higher cost for energy.
In view of the known finite fossil fuel resources,
the required compatibility between a clean environment and energy production and its use im-
poses a further element of indeterminacy on the
energy problem. The impact of the conventional
power plant's emission of pollutants into the
environment has, in the past few years, been
widely discussed and, therefore, these issues
will not be considered in this paper. The fundamental aim here will be to discuss the establishment

of a system of energy production over a relatively
short time period where all uncertainties arising
from various approaches to energy production and
conservation are minimized. Thus, deviation from
the minimum uncertainty which may lead to serious
mistakes and hence to uncontrollable global in-
stabilities may be avoided.

MINIMUM INDETERMINACY

A state of minimum uncertainty in the production and use of energy cannot be achieved with
fossil and non-fossil fuels alone. At present,
there exists a proliferation of ideas, some
feasible and some quite utopian, for the production and use of energy. One of the most important
factors is the vital role of the consumer himself
in the conservation of energy. From the point of
view of energy consumption and his resulting life-
style, the individual is a member of a statistical
ensemble contributing to the increase in entropy.
The human being's general behavior as a social,
biological, economic and technical member of this
grand ensemble is one of the basic determining
factors in the creation, as well as in the in-
hibition, of the indeterminacies associated with
the energy problem. No amount of resources,
methods of production and energy use, can lead to
a permanent and perfect solution of the energy
problem. In fact, it is now recognized that for
the rest of this century and beyond, energy will

remain a constant and fundamental survival problem of man.

Energy is the unifying and uniquely sustaining
entity of life. The world is now bracing for the
possible disaster of a global famine. This is an
energy dependent phenomena. The resource scarcity
induced by the rise in petroleum prices has
triggered a world-wide shortage of nitrogen fertilizers. As a result, farmers in under-developed
countries will have to bear the brunt of what seems
to be certain to come, i.e., starvation and
eventual deaths of millions!

Now, however, let us assume the possibility
of a photo-synthetic conversion process of sunlight to the production of proteins. Even if
oceans are used for this process as well as land
areas, we shall have to face the need for other
uses of such space which are also vital to survival. How far can man exploit a resource such
as the photo-synthetic creation of proteins without causing a detrimental effect on other nece-
ssities obtained from the used space? Thus, the
photo-synthetic conversion process of sunlight is
invariant, i.e., it always leads to the production of food, but the interference of this process
with other life sustaining uses of the same resource is indeterminate.

The natural deterioration of the environment (independent of man) can, at best, be determined within obscure limits. The same applies
to man-made pollutants introduced into the

atmosphere where, for example, we can use the
factor $\frac{E_t}{E_s}$ as an illustration of the degree of our
interference with the atmosphere. Solar radiation
transport (atomic and molecular emission and absorption of radiation in the atmosphere) through
the atmosphere is, in terms of physical, chemical
and biological phenomena, an invariant process.
However, the effects on the propagation of life
caused by such modification of the atmosphere
(i.e., $\frac{E_t}{E_s} \gg 1$) and the resulting changes of
solar spectrum observed on earth are indeterminate.
In fact, it is not even clear whether one can dis-
cover a point of minimum uncertainty in such
atmospheric interference. Do we now possess a
more tamed and hospitable atmosphere conducive to
better climatic and agricultural results than
formerly or have we inadvertently set loose a
chain of physical and chemical reactions hostile
to life itself? The accumulated results of the
"unplanned experiments" performed on the global
atmosphere since the beginning of life do not
yield predictable patterns for its future behavior.

TEMPERATURE GRADIENT

The vertical temperature gradient in conjunction with the existing ocean currents can be
used to generate electricity and this idea is,
among others, considered "reasonable" and, even,
a "feasible" method of producing usable energy.
Here, also, not only do we have to consider its

large scale effect on ocean ecology, but also its
role in the local air-sea interaction and in per-
turbations of climatic conditions.

SOLAR ENERGY

In almost all cases where a source of energy
exists the discussions of the subject are focused
on the economic feasibility, abundance, technological capability and, eventually, the question
of energy for all and for an unlimited time. This,
obviously, is quite foolhardy and credulous. The
total character of man is invariant. His wisdom
does not increase with time and is, in fact, time-
independent. It is further clear that man can
find and will, in fact, find other usable and
life sustaining energy sources but without being
in command of his own destiny!

Whenever nations face a common disaster they
communicate and cooperate with much greater ease
and understanding. The problem of energy and the
environment are fundamental examples of problems
which may initiate an era of relatively easier
(and necessary) international cooperation. From
this point of view, the energy crisis can, in
some areas, create fertile grounds for international rapproachment and peace-inducing activities. The greatest natural source of energy
is, of course, the sun itself. However, it must
be understood that the harnessing of solar
radiation for the large scale production of

electrical energy will be a new experiment in
meeting at best some of our energy needs. These
efforts must be balanced with efforts on all other
energy resources. In view of many indeterminacies
associated with energy production from various
natural sources, this global experiment will have
to be performed with the greatest care and with
most ingenious planning. At first sight, we may
expect, for all nations, great benefits from the
extensive development of solar energy: cheap
electrical energy, greater industrial producti-
vity, more food production, and fewer environmental hazards. One of the indeterminacies associated with these happy results is, aside from
environmental issues, the unknown impact it may
make on the global population growth leading to
further depletion of material resources. It is
not necessarily true that an unlimited amount of
energy is the ultimate answer to man's survival
problem. The availability of cheap electrical
energy would, of course, expedite the economic
growth of developing nations and finally expedite the rise in their standards of living;
but for how long? Perhaps a year or two! In
the absence of resources the ever increasing gap
between "have" and "have not" nations would
eventually get even wider.

There exists, at present, a variety of
proposals to convert solar radiation, directly or
indirectly, into usable electrical energy. The
greatest drawback lies in the very low value of

the solar energy density received on the earth's surface. The multiplicative role of various kinds of collectors to increase the solar energy density on a given surface area is greatly inhibited by the economic factors involved in the costly construction and maintenance of elaborate steering mechanisms as well as by the inefficiency arising from the transport of the resulting heat energy even if one heats water, for example, to drive a turbine.

In the technologically advanced countries most of the industrialized areas are located in the temperate zones which are deprived of the prolonged flow of solar energy. One of the proposed remedies for the production of electrical energy entails the collection of solar energy in outer space and its conversion to microwave form in order to minimize absorption losses during its propagation in the atmosphere. It is hoped that these microwaves can be beamed at receiving stations on earth. If we assume that all the technological hurdles are removed, from 10 to 100 million (or more) kilowatts output can be reached thus meeting the power requirements of the entire United States in 1974. At least two space stations should be placed in an orbit parallel to the earth's equatorial plane at a minimum altitude of about 36,000 km with a 22 degree phase difference to assure solar illumination for at least one of them. However the collection of radiant energy in space and its transportation

to earth via a microwave transmitter to send it
to earth is impractical, if not impossible. The
surface area dependence of the solar energy in-
tensity requires microwave collectors on the
ground about the same size as the ground level
solar energy collector. Thus, one of the in-
determinant factors in this case is the insur-
mountable economic consideration, not the technical feasibility. Even if the economic problems
are solved in terms of the present-day resources
available, in the long run we may have to face a
serious competition for these resources between
their use in "solar" electricity and other vital
uses of these materials. Finally other alternatives include direct conversion of solar energy
to electric power using cells. The economic
feasibility and problems of storage discourage an
extensive application of the method.

We shall always have to face the basic law
of nature: *Whenever man does interact with the
totality of his environment he sets free an infinite number of other natural phenomena to
counteract his specific aim to gain energy from
nature's store.*

FUSION ENERGY

It is often stated that if one of the
several promising approaches succeeds and an
energy producing thermonuclear reactor is built,
then such an energy source will remove all

resource constraints and we shall practically
have an unlimited amount of energy to use. The
validity of such claims are questionable. The
main thermonuclear fuel, deuterium (^{2}H), is
easily extractable from sea water and the amount
present is equivalent to almost a million times
the energy content of all fossil fuels. However,
the ecological, biological or other environmental
functions of ^{2}H in sea water are not known. It
is conceivable that man's interference with the
deuterium balance on a large scale may be counter-
acted by forces that do not sustain an atmosphere
favorable to life. This kind of an uncertainty
can be minimized if research is performed, along
with the fusion research, on the fundamental role
of ^{2}H in sea water pertaining to its life sustaining functions, its possible role in the interaction of the earth's atmosphere and sea.

It would take, probably, a hundred years or
more before the oceans could feel the absence of
large amounts of deuterium (nobody knows) but its
impact on the biosphere could be irreparable.
*The deuterium exhausted oceans may be partially
deprived of basic inertial properties required
in their interaction with the atmosphere. A more
immediate hazard to contemplate and to worry
about is the possibility of almost unlimited
energy at the disposal of man. The production of
large stocks of energy requires scientific and
technological ingenuity which man has always
maintained in the course of his history. However,*

*the vital issue is the use of this energy, which
requires wisdom, responsibility and discipline
of which man does not and never did have enough.*
In order to minimize the uncertainties in the
short, intermediate, and long term, we must make
use of all sources of energy. However, great care
must be taken in the estimation of the safety
problems, in the individual minimization of the
indeterminacies, in the computation of the weight
factors in each case, in maintaining a variety of
methods of energy extraction as a function of
regional requirements and regional modus vivendi.

FISSION ENERGY

There exist a variety of energy sources:
fossil fuels (oil, coal, natural gas); non-fossil
fuels (nuclear fission, nuclear fusion); solar
radiation, oceans temperature gradients; geo-
thermal power, windmills; etc. There is also an
ever increasing global demand of energy. The
usefulness of any resource of energy cannot be
measured by the amount of the available resource
and the time it will take to deplete it. For
example, *no one knows what will be the quality
of our environment when all the fossil fuels are
extracted and consumed by the vehicles of transportation, utilities, industry, agriculture, etc.*
In the case of non-fossil fuels the eventual side
effects seem to indicate more formidable difficulties and their long term cummulative impact

may span the entire human life on the earth.   The
residual ash from the thousands of nuclear reactors (conventional and breeder types) scattered
all over the globe, each producing 40,000 mega-
watts of electrical energy, presents a formidable
disposal problem.   The problem of irradiated fuel
disposal from nuclear reactors and their transportation requires greater ingenuity, more industrial and economic investment and greater risks
than the energy producing reactor itself.   Thus,
we must not endeavor to reach to a level of nuclear
energy usage where we no longer have complete
control over the eventualities and potential
hazards of this resource.   The optimistic fore-
casts on the side of ample nuclear energy, do not,
obviously, reckon with the large number of indeterminacies surrounding the nuclear potential.

There is always some danger in playing with
fire and all sources of energy do contain "fire"
in them.   Life and civilization is based on fire.
It appears that, at least at present, nuclear
fission as a source of energy is the most fiery
of all energy-yielding processes used to date.
The burning of wood, coal, oil, gas, etc., to ex-
tract energy are familiar molecular processes.
Extraction, refinement and storage of fossil fuels
are relatively simple and we have had sufficient
experience with them.   The non-fossil fuels are
new to us.   The hazards associated with nuclear
fission are unique and our experiments with them
are few compared to the potential risk.   Human

fallibility, which can never be discounted, could
be far more fatal in the case of non-fossil fuels
than what has been experienced with fossil fuels.
In fact, the point of view of this paper, based
on energy indeterminacy, considers it to be a
law of nature that no fool-proof energy producing
nuclear fission system can ever be engineered!
The latter statement applies also to fossil fuel
systems but the wounds from the failure of a
nuclear device cannot be healed. The element of
chance will unfortunately be one of the pre-
dominant features of our use of nuclear energy.
Nuclear reactors (both conventional and breeder
types) can be made to deliver energy more economically by raising the reactor power density.
The higher the power density the greater will be
the risk factors.

It appears rather naive for us to worry about
nuclear accidents in energy reactors, about the
disposition of a fast breeder's ash, or about
nuclear runaways, when the potential nuclear fire
power in the hands of governments constitute a
far greater threat to the existence of all life
itself than do the projected 10,000 or 100,000
energy-producing reactors of the world.

ENERGY, FOOD AND ENERGY BLACK HOLE

A cycle of global climatic changes is pre-
dicted to occur over the next few decades, which
will be unfavorable to agricultural production.

Thus, we may face not only an increasing demand
for food for survival, but the spectre of global
famine leading to deaths of hundreds of millions
of people. Under such extreme circumstances, de-
salination of sea water may become an absolute
necessity for growing subsistence crops. It is
estimated that it will take about 50 megajoules
of energy to desalinate one cubic meter of sea
water. On this basis we have to double the
current world energy consumption to meet a subsistence level of agriculture. However, if we
include the technological, economic, and industrial
layout necessary and the corresponding strain on
the other life sustaining uses of materials then
we may be facing a task that we never before had
to undertake for survival. This means that we may
reach a critical point where energy, usable without endangering the environment, is considerably
less than the required amount for the continuation
of life on this planet. Under these circumstances,
it would take a short time, compared to the time
spanned by the entire history of our civilization,
for life, under its own weight, to collapse.
Thus the earth would become a dead planet or an
"energy black hole". In order to avoid or to
indefinitely postpone such an eventuality we must
first recognize the global nature of the problem.
The energy interdependence of nations must not be
based on the geopolitical distribution of resources, of know-how, of technological capabili-
ties and potentials alone.

The rate of population growth is, undoubtedly, the greatest threat to survival regardless
where it occurs, in developing nations as well as
in technologically advanced nations. The fact is
that for every American, Swedish, or oil-rich
Arab baby, forty Indian or Indonesian babies are
born today and they will strain equally the resources of the world. These facts demonstrate
clearly that no nation, neither "have" nor "have
not", has any acceptable reasons for free population growth. In fact, the "zero population
growth" often advanced as an ultimate solution to
most of the frontier problems (energy, environmental deterioration, hunger, etc.) is not enough.
Mankind's chances for survival are inversely proportional to the number of people on earth and
directly proportional to the judicious and ingenious use of resources. A world-wide "negative
growth" of population from the current 4 billion
to 3 billion people must be adopted by all nations
as the ultimate goal for a state of global equi-
librium to avoid total collapse of our civilization.

The modernization of agriculture, to combat
soil deterioration and erosion, the invention of
new fertilizers for unlimited expansion of crop
production, etc., in order to feed our ever-
increasing world population would, in the long
run, be a great disservice to our world if not
associated with a negative growth of population.
The recent experience in which world grain

reserves have reached their lowest level in two
decades, equal to only about twenty-seven days
supply, does carry a significant message in regard
to the world population problem. World Organizations such as United Nations and others, established for the solution of a multitude of problems, must not have as their sole aim the reso-
lution of global frontier problems to further
sustain a level of miserable existence. The time
has come for these organizations to adopt new
principles and new goals, compatible with present
and forthcoming changes. The problem is not how
to feed, clad, provide shelter, transport, educate,
supply some of the good things of technology, to
an unboundedly growing world population. The
real problem is growth with quality which is
strongly dependent on the negative growth of
world population to a steady state level of perhaps 3 billion people. It is, certainly, preferable that three billion people exist happily and
with dignity, than it is to have the current
four billion or, with unchecked growth, the expected eight billion and beyond, live in the
"energy black hole" described above.

Thus, the real interdependence of the nations
from the point of view imposed by the greatest
problem of all time, energy, is not in the distribution of energy from the resource rich nations
to resource poor nations, but in the ultimate
increase of per capita energy consumption in a
world containing a population much less than its

current levels. This proposal is, perhaps, not a
practical one but it is the only sane path. The
alternative, namely unspared efforts to try all
possible roads to sustain a growing world with
finite resources is in the short run, a practical
approach, but its end result in the not too
distant future is the "energy black hole".

The old and still existing myth, the greater
the population of a country the better for its
defenses against foreign aggression is, of course,
no longer true. In fact, in proportion to its
resources and land, a country with smaller population could socially and economically, be more
stable than a country with a much larger population. For example, two hundred million persons
in the Indian subcontinent without an atom bomb,
would be economically a much more viable society
and militarily a much stronger one, than would a
society with six hundred million unsustainable
members in possession of atomic weapons.

The greatest aid to developing countries
with rapidly increasing population from the
"have nations" can consist of setting direct
examples in population control. The NATO countries,
as a military and political block, could begin by
initiating a negative population growth in their
own countries. In fact, because of the enormous-
ly wasteful economic structures of these countries,
their population growth is the most serious in the
world. In NATO countries the per capita strain
on environment and global resources are among the

greatest in the world. The consequences of
"negative population growth" are not known but in
case of an undesirable outcome the process can be
reversed. We may find that a pulsating global
population of between 3 and 4 billion people is
the most realistic solution to the problem of
resources, energy and environment.

PARTICIPANTS

Edward Ames
Department of Economics
State University of
 New York at Stony Brook

Joseph Aschheim
Department of Economics
George Washington
 University

M. A. B. Beg
Department of Physics
Rockefeller University

Robert Blumenthal
Department of Health,
 Education and Welfare
National Institutes of
 Health

Martin Bronfenbrenner
Department of Economics
Duke University

D. R. Caianiello
Consiglio Nazionale
 delle Ricerche
Laboratorio di
 Cibernetica

Mou-Shan Chen
Center for Theoretical
 Studies
University of Miami

Anthony Colleraine
Department of Physics
Florida State University

P. A. M. Dirac
Department of Physics
Florida State University

John Eccles
Department of Physiology
State University of
 New York at Buffalo

Erich A. Farber
Department of Mechanical
 Engineering
University of Florida

Sidney Fox
Institute for Molecular
 and Cellular Evolution
University of Miami

Nicholas Georgescu-Roegen
Department of Economics
Vanderbilt University

Donald A. Glaser
Department of Molecular
 Biology
University of California
 at Berkeley

Melvin Gottlieb
Plasma Physics Laboratory
Princeton University

Gary Higgins
Lawrence Livermore
 Laboratory
University of California

Joseph Hubbard
Center for Theoretical
 Studies
University of Miami

C. S. Hui
Center for Theoretical
 Studies
University of Miami

Henry Hurwitz
General Electric Company
Schenectady, New York

Abraham Klein
Department of Physics
University of
  Pennsylvania

Behram Kursunoglu
Center for Theoretical
  Studies
University of Miami

Willis E. Lamb, Jr.
Physics Department
Yale University

Joseph Lannutti
Department of Physics
Florida State University

Sidney Meshkov
Radiation Theory Section
National Bureau of
  Standards

Stephan L. Mintz
Center for Theoretical
  Studies
University of Miami

Laurence Mittag
Center for Theoretical
  Studies
University of Miami

Lars Onsager
Center for Theoretical
  Studies
University of Miami

Edwin E. Salpeter
Laboratory of Nuclear
  Studies
Cornell University

Julian Schwinger
Department of Physics
University of California
  at Los Angeles

George Soukup
Center for Theoretical
  Studies
University of Miami

Ichiji Tasaki
National Institute of
  Mental Health
Laboratory of
  Neurobiology

Edward Teller
Lawrence Berkeley
  Laboratory
University of California

Georges Ungar
Department of
  Anesthesiology and
  Pharmacology
Baylor College of
  Medicine

Jan Peter Wogart
Institute of Inter-
  American Studies of
  the Center for
  Advanced International
  Studies
University of Miami

OBSERVERS:

George Adelman
Managing Editor and
  Librarian
Neurosciences Research
  Program
Massachusetts Institute
  of Technology

Robert Lind
Department of Physics
Florida State University

F. David Peat
National Research Council
  of Canada

SUBJECT INDEX